Money After 40

Closing the Gap When Retirement Feels Too Close

Joseph Trek

Published by NobleTrex Press

For permissions and other inquiries, write to:

P.O. Box 3132, Framingham, MA 01701, USA

Contents

1 The Late Starter's Reality Check **11**

1.1 Confronting the Gap 11

1.2 Defining Your "Enough" 19

1.3 The Truth About Your Net Worth 29

1.4 Cash Flow: The Pulse of Your Plan 39

1.5 The late starter's advantage lives in cash flow . . 40

1.6 Why budgets fail (and how to do one that won't) 41

1.7 Income: count it the way your life receives it . . . 42

1.8 Expenses: stop arguing about morality and start sorting by behavior 43

1.9 A concrete case: Sam's "I make good money, where is it going?" moment 43

1.10 The most important calculation: your monthly surplus 45

1.11 How to route the surplus (the order that keeps you calm) 46

1.12 Automation: the late starter's best friend 46

1.13 A reasonableness check (without comparing your life to strangers) 47

1.14 The "Future You" pipe is a lever, not a leftover . 48

1.15 Actionable takeaway 48
1.16 Rewiring Your Money Mindset 49

2 Creating Margin: Cash Flow, Debt, and Income 59
2.1 Budgeting for the Time-Crunched 59
2.2 Planning for the Unexpected 70
2.3 Automating Your Catch-Up Plan 80
2.4 Debt Triage and Credit Health 88
2.5 Strategic Debt Elimination 98
2.6 The Income Accelerator 106

3 Securing the Foundation: Safety Nets and Housing 117
3.1 Building a Bulletproof Emergency Fund 117
3.2 Insurance: Protecting Your Earning Power 130
3.3 Life Insurance for Midlife Families 142
3.4 The Housing Dilemma: Rent, Buy, or Hold? . . . 153
3.5 Downsizing as a Wealth Accelerator 165
3.6 Estate Planning Essentials 175

4 Investing for Growth: The Catch-Up Engine 185
4.1 Compounding on a Deadline 185
4.2 Risk, Return, and the Fear of Loss 195
4.3 Asset Classes Decoded 205
4.4 The Power of Diversification 214

4.5 Low-Cost Indexing Strategies 223
4.6 Designing Your Asset Allocation 233
4.7 Staying on Track: Rebalancing 242

5 Supercharging Savings with Tax Advantages 253
5.1 Tax Basics for Late Savers 253
5.2 Mastering the Workplace Plan 264
5.3 Unlocking Catch-Up Contributions 275
5.4 IRAs: Flexibility for Your Future 285
5.5 Consolidating Your Financial History 294
5.6 Tax-Efficient Investing Habits 305

6 The Finish Line: Planning Your Retirement 317
6.1 Estimating Your Real Retirement Number 317
6.2 Why the 80% rule keeps missing the point 318
6.3 The three-phase lens: Go-Go, Slow-Go, No-Go . 319
6.4 Start with a bottom-up budget (yes, even if you hate budgeting) 320
6.5 The expenses that "disappear" (and the ones that don't) 321
6.6 A concrete case: Maria and Devin, age 47 and 50 322
6.7 Convert spending into a portfolio target (without pretending the future is certain) 325
6.8 Make room for taxes (the stealth expense) 326

6.9 The "one-time" expenses that deserve their own bucket . . . 327
6.10 Retirement planning gets easier when you name the life you want . . . 328
6.11 Actionable takeaway: build a "Retirement Budget in One Hour" draft . . . 329
6.12 The Healthcare Planning Gap . . . 330
6.13 Social Security Optimization . . . 339
6.14 Measuring the Gap: Scenario Planning . . . 349

7 Converting Savings to Income 361

7.1 Surviving the "Red Zone": Sequence of Returns Risk . . . 361
7.2 Engineering Your Paycheck: Withdrawal Strategies That Flex . . . 374
7.3 The Tax Torpedo: Sequencing Your Withdrawals 384
7.4 Three flavors of money, three kinds of tax consequences . . . 385
7.5 Tripwire #1: Social Security taxation (it's not all-or-nothing) . . . 386
7.6 Tripwire #2: Medicare IRMAA (the cliff you don't feel until two years later) . . . 387
7.7 Tripwire #3: Required minimum distributions (RMDs) turn "maybe" into "must" . . . 388
7.8 The comforting rule of thumb—and the reason it sometimes backfires . . . 389

7.9 A concrete case study: the couple who "did everything right" . . . 390
7.10 A practical sequencing framework that respects the tripwires . . . 391
7.11 1) Cover the basics with the least "tax collateral damage" . . . 392
7.12 2) Use tax-deferred money on purpose (not by default) . . . 392
7.13 3) Treat the Roth IRA as your tax and IRMAA shock absorber . . . 393
7.14 The "gap years": the most underused opportunity in retirement tax planning . . . 393
7.15 Actionable takeaway: build a one-page "income map" before you take another withdrawal . . . 394

8 The Execution Toolkit **397**

8.1 Your Digital Financial Dashboard . . . 397
8.2 Protecting Your Wealth from Fraud . . . 410
8.3 Building Your Financial Team . . . 420
8.4 The first rule: stop shopping for a person, start shopping for a job . . . 421
8.5 When DIY is smart (and when it becomes expensive pride) . . . 422
8.6 The core distinction: fiduciary vs. salesperson . . 423
8.7 How to spot incentives in the wild (without becoming cynical) . . . 424

8.8 The team you actually need (and the moments you need them) 425

8.9 A case study: the advisor who "found" $90,000 (without higher returns) 426

8.10 How to verify a professional (without trusting Google reviews) 428

8.11 The compensation menu (and what you're really paying for) 429

8.12 The interview questions that reveal everything . 431

8.13 Red flags that deserve an immediate exit 432

8.14 How to use a professional without outsourcing your life 432

8.15 Actionable takeaway: write a one-paragraph "job posting" for your money 433

8.16 The Road Ahead 434

Introduction

There is a specific type of chart that haunts every finance presentation, every retirement seminar, and every "Investing 101" blog post. You know the one.

It usually features two characters. Let's call them "Early Earl" and "Late Linda." Earl starts saving $200 a month at age 20. He stops saving entirely at age 30. Linda waits until she is 40 to start. She saves the same amount every month until she is 65. The chart then triumphantly reveals that Earl, the virtuous 20-year-old, retires with significantly more money than Linda, despite Linda saving for 25 years and Earl only saving for 10.

The math is irrefutable. Compound interest is a miracle.

But if you are reading this, that chart isn't inspirational. It's infuriating. It feels less like a financial lesson and more like a moral indictment. It tells you that the defining moment of your financial life happened two decades ago, while you were busy trying to figure out your career, pay off student loans, or simply pay rent. It suggests that because you missed the starting gun, the race is already over.

I am here to tell you that the chart is lying.

Well, the math isn't lying, but the story is. That chart assumes that your income, your discipline, and your capacity to save are static. It assumes that "Late Linda" is the same person at 40 that she was at 20, just with more wrinkles.

That is the flaw in almost all standard financial advice. It fetishizes *time* while ignoring *power*.

When you are 20, you have time, but you have no power. You are likely earning an entry-level salary, you have zero assets, and your "disposable" income is usually the price of a Friday night pizza.

When you are over 40, the equation flips. You have less time, yes. But you have significantly more power. You are likely in your peak earning years. You know who you are. You know what you value. You have professional skills, perhaps some home equity, and-crucially-a sense of urgency that a 25-year-old cannot possibly manufacture.

This book is about leveraging that power. It is about discarding the regret of what you didn't do in the past and ruthlessly optimizing what you can do right now. We are going to close the gap.

The Myth of "Too Late"

Let's address the elephant in the room immediately: the panic.

Somewhere around your 40th, 45th, or 50th birthday, the fog of early adulthood lifts, and the horizon of retirement suddenly snaps into sharp focus. For years, "retirement" was an abstract concept, a hazy vacation that would happen to Future You. Now, it has a shape. It has a deadline. And when you look at your 401(k) balance or your savings account, the number staring back at you creates a physical sensation in your chest.

You run the generic online calculators, and they spit out terrifying numbers. They tell you that to maintain your current lifestyle, you need $2.5 million, and based on your current trajectory, you will arrive at age 65 with a fraction of that.

The natural human reaction to this data is paralysis. We feel shame. We feel stupid. We ask ourselves where the money went. *Was it the coffees? The vacations? The car lease?*

We dwell on the "sunk costs" of our lives. In economics, a sunk cost is money that has already been spent and cannot be recovered. Rational decision-making requires you to ignore sunk costs. You cannot un-buy that house in 2007 right before the crash. You cannot un-marry the spouse who took half your net worth in the divorce. You cannot refund the tuition for the degree you aren't using.

The only assets you can manage are the ones you have today: your current capital, your current income, and the years remaining until you choose to stop working.

Here is the good news-and I mean this literally, not just as a motivational platitude. The math of "catching up" works differently than the math of "starting early." When you have a 40-year horizon, your greatest lever is the rate of return (compound interest). When you have a 15 or 20-year horizon, your greatest lever is your *savings rate.*

This is a critical distinction. You cannot control the stock market's returns. But you have significantly more control over the gap between what you earn and what you spend. Because you are likely earning more now than ever before, every percentage point of spending you cut, and every raise you bank, acts as a turbocharger for your net worth.

We are not trying to drift slowly toward a finish line anymore. We are building a drag racer. We are stripping out the weight, upgrading the engine, and flooring the accelerator. It's a different sport entirely.

The Architecture of the Catch-Up

Most finance books follow a gentle, meandering path. They encourage you to cut out lattes and wait 40 years. We don't have that luxury. We need an aggressive, structural approach to wealth building. We need to perform triage on your finances.

This begins not with a budget app, but with a reality check. We have to look at the numbers without flinching. This means calculating your true net worth-assets minus liabilities-and staring it in the face. It's uncomfortable. It might hurt. But it is the only solid ground upon which we can build. We have to strip away the vague anxiety of "I'm behind" and replace it with the concrete data of "I need to generate $X,000 per month."

Once we have the baseline, we have to find the fuel. This is where most midlife financial plans fail. They assume you can just "save more" without acknowledging the massive friction of midlife expenses. You might be supporting aging parents and college-bound children simultaneously. You might be paying for a mortgage in a high-cost-of-living area.

We have to create *margin*. This isn't about clipping coupons; it's about structural changes. We will look at the big levers: your housing costs, your debt load, and your income.

Housing, in particular, is often the sacred cow of midlife finance. We cling to the family home because it represents stability. But for many late starters, a large, equity-rich but cash-poor house is an anchor dragging them down. We will have the hard conversation about whether your zip code is keeping you poor, and how unlocking that equity could instantly bridge years of missed savings.

Then, we turn to the income side of the ledger. In your 20s, "earning more" meant hoping for a promotion. In your 40s and 50s, you have leverage. You have a network. You have expertise. We will explore how to negotiate, how to pivot, or how to create a side income stream that goes 100% into your catch-up fund. If you can earn an extra $20,000 a year and invest it all, that isn't just $20,000; that is potentially $300,000 or more in your retirement nest egg over 15 years.

The Investing Engine

Once we have squeezed cash out of your lifestyle and boosted your income, where does it go?

If you are risk-averse, you might be tempted to keep it in cash, terrified of a market crash right before you retire. If you are panic-stricken, you might be tempted to gamble on speculative assets-crypto, options, the "next big thing"-hoping for a lottery ticket to save you.

Both are disasters.

We need a sophisticated investing strategy that respects your shorter timeline. We need growth, but we cannot afford a catastrophic loss that takes a decade to recover from. This requires a specific type of asset allocation-a blend of stocks for growth and bonds or alternatives for stability.

We also have a secret weapon that 25-year-olds don't utilize fully: the tax code. The government knows you are behind. That is why "Catch-Up Contributions" exist. The IRS literally gives you extra space to shovel money into 401(k)s and IRAs once you turn 50. We are going to exploit every single loop, tax break, and deduction available to ensure that your money is growing for *you*, not for the Treasury.

Redefining the Finish Line

Perhaps the most important shift we will make is redefining what "retirement" actually means.

The old model-work until 65, receive a gold watch, and do nothing but play golf until you die-is obsolete. It's also incredibly expensive to fund.

For the late starter, the goal is often "Financial Independence," or perhaps a hybrid retirement. It might mean saving enough to leave the high-stress corporate job at 60 and coasting with a lower-stress consulting gig that covers your expenses while your nest egg continues to compound in the background.

We will run the scenarios. We will look at Social Security not as a boring government handout, but as a longevity annuity that you need to optimize strategically. We will look at healthcare, the silent wealth killer, and plan for the gap before Medicare kicks in.

We will replace the vague terror of the unknown with a series of solvable problems.

The Psychology of the Second Half

There is a psychological phenomenon known as the "U-curve of happiness." Research consistently shows that human happiness tends to dip in midlife-bottoming out in our late 40s-before rising again as we enter our 50s and 60s.

The bottom of that curve is where the financial anxiety hits hardest. It is the "midlife crisis" zone. But it is also the turning point.

When you decide to take control of your money after 40, you are doing something profound. You are signaling to yourself that your future matters. You are breaking the cycle of avoidance.

I have worked with hundreds of people who woke up at 45, 50, or even 55 with almost nothing saved. I have seen them turn it around. I have seen them pay off six figures of debt. I have seen them downsize homes they thought they couldn't live without, only to find they were happier with less stuff and more freedom.

They didn't do it by finding a magic stock pick. They did it by getting serious. They treated their personal finances like a business turnaround project. They cut the bleeding, streamlined operations, and maximized revenue.

You have lived through the experimental phase of your life. You have the scars and the wisdom to prove it. Now, you are entering the execution phase.

A Note on "Experts"

As we move through this material, you will notice I don't speak much in absolute rules. Financial gurus love absolute rules. *"Never buy a latte." "Always pay off the mortgage." "Never use credit cards."*

Rules are for people who don't understand mechanics. We are going to learn mechanics.

Mechanics allow you to understand *why* you are doing something. Should you pay off your mortgage or invest that extra $1,000? A rule says "do this." Mechanics say, "Compare the after-tax interest rate of the debt to the expected after-tax return of the portfolio, adjusted for your risk tolerance."

Don't worry, we'll make the math easy. But I want you to understand the *why*, because when things get tough-and they will-rules shatter. Understanding survives.

You are going to encounter friends and colleagues who are on different paths. You will see the neighbors with the new Teslas and the renovations. You might feel a pang of jealousy. But you have to remember: you don't see their balance sheets. You don't see their credit card statements. You are playing a different game now. You are playing for freedom, not for show.

How to Use This Book

This is not a novel. You don't have to read it linearly, though the logic flows from mindset to mechanics to execution.

If you are currently drowning in credit card debt, jump to the section on *Debt Triage*. If you have a pile of cash but are paralyzed by fear of the stock market, go to *Investing for Growth*. If you are worried about taxes eating your catch-up contributions, head to *Supercharging Savings*.

However, I strongly recommend you read the first chapter, *The Late Starter's Reality Check*, before doing anything else. It establishes the "Enough" metric. If you don't know what "enough" looks like, you will run a race with no finish line, and you will exhaust yourself.

The Best Time to Plant a Tree

You know the old Chinese proverb: "The best time to plant a tree was 20 years ago. The second best time is today."

It's a cliché because it's true. But I prefer a different analogy.

Imagine you are on a flight from New York to London. You are halfway across the Atlantic. Suddenly, the pilot announces that due to a headwind, you are behind schedule.

Does the pilot turn the plane around? Does the pilot give up and land in the ocean?

No. The pilot checks the fuel reserves. The pilot calculates the necessary speed adjustments. The pilot looks for a more favorable altitude with better winds. The pilot focuses on the controls.

You are the pilot. You are halfway across the ocean. You might be behind schedule, but you have fuel, you have altitude, and you have a destination.

We are not turning around. We are not landing in the ocean. We are going to adjust the throttle, check the instruments, and land this plane safely.

Let's get to work.

Chapter 1

The Late Starter's Reality Check

Turning 40 often triggers a specific kind of 3 a.m. panic: the sudden realization that the 'someday' you were saving for is rapidly becoming 'today.' But panic is a poor financial planner. In this chapter, we hit the pause button on the shame spiral to take a brutally honest, yet hopeful, look at where you actually stand. We aren't just checking boxes; we are locating your GPS coordinates so we can plot a realistic route to a retirement that fits your life, not a generic roadmap you aged out of a decade ago.

1.1 Confronting the Gap

At some point after 40, the retirement conversation changes tone. In your 20s, it's all *shoulds*: you *should* start early, you *should* invest regularly, you *should* let compounding do the heavy lifting. After 40, it becomes a quieter, heavier sentence: *I should have already done this.* It's not just math. It's the feeling that time has started making demands.

That anxiety has a particular flavor. It's not the sharp panic of an emergency. It's the low-grade dread that arrives when you open a retirement account for the first time in a year (or five), see a number that looks too small, and do the mental arithmetic you *didn't* want to do. You imagine the finish line moving away as you run. Then your brain does what brains do when a situation feels uncomfortable and complex: it tries to protect you by changing the subject. Scroll. Snack. Work. Anything except the one thing you actually need: a clear-eyed look.

The cruelest part is how quickly late-starting gets moralized. You're not just behind; you're *bad with money*. You're not just missing years of saving; you're *irresponsible*. That story is both common and useless. People start late for a hundred reasons that have nothing to do with character: student loans, divorce, immigration, caregiving, layoffs, chronic illness, a business that took longer than expected to stabilize, years where the goal was simply to keep the lights on. Even the people who look "ahead" often got there through circumstances they didn't manufacture. The point isn't to relitigate the past. The point is to stop using it as a reason to freeze.

Here's the reality check that stings and then sets you free: yes, you missed some compounding years. That matters. But compounding is not a *virtue reward* for people who started early. It's a tool. And the other tool—the one late starters often have more of—is *intensity*. In plain language: you may have fewer years, but you can often put more dollars to work per year than you could at 25. You've likely built skills, credibility, and earning power. You can negotiate. You can cut expenses without it feeling like a dorm-room diet. You can make decisions you couldn't make when everything was fragile.

Intensity doesn't mean punishing yourself. It means accepting a simple trade: if you want a retirement that isn't defined by anxiety, you need to redirect a meaningful amount of your current income toward the future. Not someday. Now. And the most important mental shift is this: you are not trying to "catch up to other people." You are trying to build *your own* runway before the calendar builds it for you.

A lot of advice fails late starters because it treats time as the only lever. "Start early" is great advice—for someone who can still do it. Late starters need a different frame: retirement is not a trophy for people who did everything right from age 22 onward. Retirement is the result of cash flow decisions made consistently enough for long enough. Those decisions can be made at 40, 45, even 55. The tactics change. The seriousness changes. The opportunity is still real.

One of the most revealing exercises is to stop thinking in decades and start thinking in paychecks. If you get paid twice a month, that's 24 chances per year to make a future-friendly choice. Over 15 years, that's 360 decisions. Most people imagine retirement savings as a single heroic act: *I need to save a mountain of money.* In practice, retirement is built by repeating a smaller act hundreds of times: *Before I spend this, I reserve a slice for future me.* When you frame it that way, the gap stops being a single terrifying number and becomes a set of controllable actions.

There's also a reason late starters feel uniquely haunted by what they *didn't* do: you can now see the consequences. When you were younger, the future was abstract. At 45, you can picture your 65-year-old self with an almost uncomfortable clarity. You can feel the strain of supporting kids, parents, and a mortgage all at once. You can sense that your career isn't infinite. The discomfort is not a sign that you're failing; it's a sign that your brain is finally treating the future as real.

Now for the part that surprises people: starting late can be an advantage in one specific way. When you start early, it's easy to be casual. You can toss in a few percent, miss a year, shrug, and assume time will cover your sins. Starting late forces you into seriousness. It pushes you toward higher savings rates, cleaner systems, and better attention to taxes and fees. In other words, it doesn't feel good, but it can make you more effective. The goal is not to be *perfect*. The goal is to be *deliberate*.

Consider a real-world example that plays out quietly in countless households.

Denise is 42. She works in healthcare administration. Her career finally stabilized in her late 30s after a messy divorce and a few years of underemployment. She has $28,000 in a 401(k), $6,000 in a Roth IRA she opened and forgot, and $9,000 on a credit card from the divorce transition. She earns $112,000. Her rent just jumped. She tells herself she's "bad at money," which conveniently excuses her from looking closely.

Denise is not an edge case. She is normal. What changes her trajectory isn't a sudden windfall. It's a two-part decision: (1) she stops using the account balances as a referendum on her intelligence, and (2) she turns her higher earning years into a savings engine.

First, she faces a hard truth with a soft grip: the money she already saved is not the headline. The headline is what her *cash flow* can do now. She runs her numbers and realizes she can redirect $1,800 per month without wrecking her life if she makes three changes: she refinances an auto loan, cancels two subscriptions she forgot she had, and—the big one—moves to a slightly smaller apartment when her lease ends. None of these are "fun." But none are permanent. They're a 3-year sprint.

Next, she uses a structure that doesn't rely on willpower. She increases her workplace retirement contribution on payday, so the money never lands in her checking account. She sets a separate automatic transfer for her credit card payoff. She gives herself a modest, pre-approved spending amount each week—not because she's a child, but because decision fatigue is real and her life is busy. She doesn't try to become a new person overnight. She builds rails that keep the train moving even when she's tired.

After 18 months, two things happen that most people don't expect. First, the anxiety drops. Not because she has "enough" yet, but because she is no longer improvising. Second, her identity shifts. She stops seeing herself as someone who is behind and starts seeing herself as someone who is executing. That identity is powerful. It tends to keep paying dividends.

Denise's story illustrates the late starter's advantage: intensity plus structure. The math still matters, but the psychology is what gets the math done.

Let's talk plainly about the gap itself. The gap is not just the difference between what you have and what you want. It's the difference between your current trajectory and your required trajectory. If you keep doing what you're doing, where do you land? If you change one lever—savings rate, retirement age, investment risk, spending level—how does that landing point move? Late starters often assume the gap is fixed, like a canyon. It's more like a sliding scale. You can move it.

There are four levers that matter most, and they don't all require you to "save more."

First: savings rate. This is the obvious one, and it's powerful. If you're starting late, a 5% savings rate is usually a polite suggestion, not a plan. Late starters commonly need 15%–25% of income saved toward retirement, depending on goals and starting point. That sounds brutal until you remember: you're not trying to live like a monk. You're trying to redirect money you're currently spending on things that do not meaningfully improve your life.

Second: spending level in retirement. The gap shrinks dramatically when "retirement" is defined as *freedom from forced work,* not *a permanent vacation financed by a giant portfolio*. Many people don't need a luxury number. They need safety, healthcare coverage, and the ability to say no to nonsense. When you define your target lifestyle clearly, the goal becomes more realistic.

Third: retirement age (or semi-retirement). One extra year of work can be surprisingly valuable. It's another year of contributions, another year for investments to grow, and one fewer year of withdrawals. Sometimes the best "catch-up" move is not saving an additional $10,000 per year; it's designing your career so you can work two or three years longer without hating your life.

Fourth: claiming strategy for Social Security. This is a big deal for late starters because it can function like inflation-adjusted lifetime income. The decision of when to claim can permanently change your monthly benefit. You don't need to memorize rules right now. You do need to treat the decision as a lever you'll eventually pull on purpose, not by default because you're tired or scared.

Notice what's missing from that list: shame. Shame feels like urgency, but it behaves like sedation. It makes people avoid the very actions that produce change.

There's another common late-starter trap: trying to "make up for lost time" with dramatic investing. People look at the gap and think, *I need something that doubles.* They chase hot stocks, cryptocurrency, complicated options strategies, or expensive "alternative" investments pitched as sophisticated. Sometimes they get lucky. Often they just add volatility to a plan that already has less time to recover from mistakes.

Intensity should show up in your *savings behavior*, not in your risk-taking. Late starters benefit from boring, consistent investing more than almost anyone, because boring is repeatable. Boring doesn't require you to be right about the future. Boring lets you focus on what you control: how much you invest, how often, and how cheaply.

If you want a mindset that actually helps, borrow this line: *My job is not to win. My job is to not lose.* Not lose to taxes you could have avoided. Not lose to fees you didn't notice. Not lose to lifestyle creep disguised as "I deserve it." Not lose to procrastination that quietly steals the most valuable asset you have left: time.

A surprisingly effective move is to convert abstract time into a deadline you can feel. Pick a date when you want the option to stop full-time work: 60, 62, 65, 67. Then count the months.

If you're 45 and you want the option at 65, that's 240 months. That number does two things. It makes the situation real, and it makes it workable. Two hundred forty months is not a foggy life sentence. It's a countdown with steps. You can improve your situation meaningfully in 12 months, and the proof will show up in your statements.

Late starters also tend to underestimate how quickly behavior becomes normal. The first month of a higher savings rate feels like wearing shoes that don't fit. The third month feels inconvenient. By month six, it's the new baseline. The biggest win isn't the dollars; it's the fact that you stop negotiating with yourself every week. You become the kind of person who saves before spending. That identity is worth more than a spreadsheet.

There's a practical reason to confront the gap early, even if it's uncomfortable: small course corrections made now prevent desperate ones later. When people avoid reality for too long, the only remaining fixes are painful: withdrawing retirement money early, taking Social Security as soon as possible, carrying high-interest debt into their 60s, or working longer than their health allows. Those outcomes aren't inevitable, but avoidance makes them more likely. The earlier you face the numbers, the more options you keep.

And options are the real prize. Not bragging rights. Not a certain account balance screenshot. Options: the ability to walk away from a toxic boss, to reduce hours when your parents need care, to handle a health curveball without financial freefall, to choose work that suits you rather than work that panics you.

If you're tempted to think, *But I'm so far behind,* try this reframe that works because it's true: your plan doesn't require you to become wealthy. It requires you to become consistent. Consistency plus time plus reasonable investing can produce surprising outcomes, especially when your contributions are meaningful.

So here's the late starter's reality check, delivered with respect: you can't go back and start at 25. You also don't need to. You need a system that turns today's earning power into tomorrow's freedom, and you need it to run whether you feel motivated or not.

Actionable takeaway: within the next 48 hours, do one small action that creates intensity without drama: increase your workplace retirement contribution by 1% (or by $50 per paycheck), scheduled to start on your next payday. If that feels too easy, do it twice: increase it now, and schedule another 1% increase to happen automatically three months from now. The point isn't the percentage; it's proving to yourself that the gap responds to action. Once you see it move, you stop arguing with the calendar and start building leverage.

1.2 Defining Your ``Enough''

The fastest way to make retirement planning feel impossible is to start with someone else's number. The internet loves a headline figure: *You need $3 million.* Or *$5 million.* Or the evergreen favorite, *a million dollars* said with the same tone people use for "a unicorn." Those numbers land like a brick because they're not really advice. They're entertainment. They're also often a mash-up of assumptions that may have nothing to do with your life: a big-city housing budget, a long retirement, high healthcare costs, generous travel, and a portfolio designed to survive almost anything.

The more useful question is smaller and more personal: what does *enough* mean for you?

Not *rich*. Not *comfortable compared to the neighbors*. Enough. Enough to sleep. Enough to say no. Enough to handle the boring emergencies without financial whiplash. Enough to avoid becoming a reluctant employee of your own fear.

And here's the twist: once you define enough clearly, retirement planning becomes less like staring up at a cliff and more like pricing a life. You're no longer guessing at a single monster number. You're building a bill.

The retirement number is a story you tell your money

Most people carry a hidden script that goes like this: "Retirement means I stop working, but nothing else changes, except I'll probably travel more, and healthcare will be expensive, and inflation is scary, and I don't want to be a burden." That script produces a number that feels both huge and vague. Vague numbers are especially dangerous because they invite two equally bad responses: you either give up because it feels unattainable, or you over-save out of fear and accidentally sacrifice the present for a future that may not need that level of funding.

The antidote is specificity.

You're going to build *your* enough number from the ground up, using three layers that reflect how people actually live:

- *Survival*: the non-negotiables that keep your life stable and safe.
- *Comfort*: the things that make life pleasant and sustainable.
- *Luxury*: the extras that are delightful, optional, and often seasonal.

A late starter's secret weapon is that you don't need to fund all three layers on day one. You need to fund survival first, then add comfort intentionally, then treat luxury as a dial you can turn up or down depending on market returns, health, and the kind of life you actually want at 70.

Start with a ``survival" budget that respects reality

Survival is not ramen noodles and suffering. It's the baseline that prevents regret. It includes housing, food, utilities, transportation, insurance, basic healthcare, and the small set of subscriptions and services that keep your life functioning (phone, internet, maybe a modest streaming plan if that's your cheap entertainment).

The quickest way to build this number is to use your current spending as raw material, then edit it for retirement. Retirement changes the shape of expenses, not just the total. Commuting might drop. Work clothes and lunches out might drop. Healthcare premiums and out-of-pocket costs might rise. Taxes might change. Your mortgage might be gone—or it might not. The point is to stop pretending retirement is a magic switch that turns spending into a different species.

A practical method:

1. Take your current monthly spending and sort it into three bins: *must-pay, nice-to-have, leaks.*
2. Convert must-pay into your survival number.
3. Decide what "comfort" means in plain English (not categories) and add those line items one by one.

"Leaks" aren't morally bad. They're the costs you wouldn't defend if you had to pay for them twice. They're the forgotten apps, the shopping that happens when you're tired, the delivery fees that quietly become a second grocery bill. If you're trying to define enough, leaks matter because they inflate your estimate of what life truly costs.

A concrete case: Maria and James, and the myth of the $3 million target

Maria (46) and James (48) live in a mid-sized U.S. city. They have two teenagers, a mortgage, and careers that finally pay well after years of climbing. Maria is a project manager earning $135,000. James is in sales and averages $110,000 with commissions. They have $210,000 in retirement accounts combined, $30,000 in cash, and a mortgage balance of $240,000. They keep hearing variations of "you need millions" and it makes them swing between panic-saving and avoidance-spending.

So they do a different exercise. They build their retirement spending "bill" as if they were pricing a life they'd actually want.

They start with survival:

- Housing (property taxes, insurance, maintenance, utilities): $2,250/month
- Food: $700/month
- Transportation (one car, insurance, gas, maintenance): $650/month
- Healthcare premiums/out-of-pocket estimate: $900/month
- Phones/internet: $180/month
- Misc essentials (basic clothing, home supplies): $300/month

That's $4,980/month, or about $59,760/year.

Then comfort:

- Eating out and simple fun: $350/month
- Gifts/holidays: $200/month
- Modest travel (one trip plus a few weekends): $500/month
- Hobbies/gym: $150/month

That's another $1,200/month, or $14,400/year.

Their *comfortable enough* estimate lands around $74,160/year (call it $75,000).

Now something important happens: the number stops being scary and starts being *useful*. They can have a real conversation. Maria realizes she doesn't actually want luxury travel every year; she wants time. James realizes he does want travel, but he'd happily trade some restaurants for more trips. They're not debating a fantasy retirement. They're negotiating an actual future.

Then they ask the question that turns spending into a target: what income sources will cover that annual number?

They expect some Social Security eventually. (They don't guess wildly; they pull estimates from their statements later, but for now they keep it conservative.) They assume, for planning purposes, that Social Security might cover $35,000—$45,000 per year for the household depending on claiming age and earnings history. That means their portfolio might need to cover roughly $30,000—$40,000 per year to hit their $75,000 target.

Suddenly, the "$3 million or bust" narrative loses its grip. If their gap is $35,000/year, a rough rule of thumb often used for initial planning is to multiply by about 25 (the idea behind a 4% starting withdrawal rate). That suggests a portfolio target around:

$$\$35{,}000 \times 25 = \$875{,}000$$

Even if they choose to be more conservative—say, $\$35{,}000 \times 28$ or $\$35{,}000 \times 30$—they're still not automatically in $3,000,000 territory. Their actual number might end up higher or lower depending on taxes, healthcare, inflation, market returns, and how long they live, but the point is powerful: *their life does not require the internet's number.*

This is what defining enough does. It replaces fear with design.

The ``enough'' number is two numbers, not one

Most retirement calculators spit out one target: *You need $X.* Real life is messier and kinder than that. You need two numbers:

- *The floor*: the survival number—what must be covered even in a bad market year.
- *The comfortable target*: survival plus the things that make retirement emotionally sustainable.

The floor is what you protect. The comfortable target is what you aim for.

This matters because it changes how you react to market volatility later. When markets drop, people who planned with a single rigid number tend to panic. People who know their floor can say, "We can reduce the comfort layer this year and still be okay." That's not deprivation. That's resilience.

A word about ``rules of thumb'' (and why they still help)

You'll hear a lot about the "4% rule." It's not a law of nature. It's a starting framework: if you withdraw about 4% of your portfolio in the first year of retirement and adjust for inflation after that, historical data suggests the portfolio had a good chance of lasting about 30 years under many market scenarios. The problem is that people treat it like a promise.

Late starters should treat rules of thumb the way you treat a weather forecast. It doesn't guarantee sunshine, but it tells you whether to pack a jacket.

Use the 4% idea to translate annual income needs into a rough portfolio target, then pressure-test it:

- What if you retire earlier than planned?
- What if healthcare costs are higher?
- What if markets are mediocre for a decade?
- What if one of you lives to 95?

If those questions feel overwhelming, good. They should. Not because you're doomed, but because they force you back to the most powerful move you can make: defining what is non-negotiable, and what is adjustable.

The overlooked lever: time and claiming decisions

For late starters, "enough" isn't just about how much you accumulate. It's also about the shape of your income.

Social Security is a major part of that shape. Claiming earlier typically means a smaller monthly check for life; claiming later typically means a bigger one. People often claim early because they feel behind, but feeling behind is not always the same as being unable to wait. Sometimes the best "investment" available to you is designing a plan that lets you delay claiming, because a higher guaranteed monthly benefit can reduce the portfolio income you need to generate forever.

You don't need to optimize this today. You do need to recognize it as part of the enough conversation. If you define enough as "$75k per year," and you can increase your reliable lifetime income, you've just reduced the amount your investments must cover. That makes everything else easier.

Enough should include dignity costs

Here's a category people forget until it hurts: the cost of dignity. Dignity costs are the expenses that protect you from desperation: a reliable car (or ride service budget) so you can get to appointments, a home environment that's safe, dental work you can afford, a little buffer so you don't have to ask your kids for help with every surprise bill, the ability to hire assistance if your body stops cooperating.

Dignity is not luxury. It's stability. It belongs in survival.

And if this is where your anxiety spikes, it's also where defining enough becomes emotionally meaningful. You're not saving for a number. You're saving to keep your future self from being cornered.

Turning your values into line items (without turning life into a spreadsheet)

Values talk can get mushy fast, so here's a clean translation: values are what you refuse to regret.

If you care about family, you might budget for being able to travel to see grandkids or to attend weddings without debt. If you care about independence, you might prioritize a paid-off home or a larger cash buffer. If you care about community, you might want room in the budget for giving or volunteering without needing a side hustle to cover basics.

When values become line items, the plan gets easier to follow. You're no longer "saving more." You're buying something specific: *the ability to say yes to the life you actually want, and no to the life you fear.*

A helpful question: if your retirement account could only buy you three things, what would they be? Not categories like "travel" or "freedom." Actual scenes: hosting a holiday meal without stress, taking a two-week trip once a year, covering health costs without bargaining with yourself. Those scenes become your comfort layer. Everything else is optional.

Make the first version of ``enough'' slightly wrong on purpose

Perfection is a delay tactic in a tuxedo.

Your first enough number will be wrong because you don't know everything yet: taxes, future healthcare policy, interest rates, the exact cost of your future home. Waiting for certainty is how people accidentally choose the default retirement plan: worry.

Build a first-draft enough number that is directionally right, then revise it once a year. You don't need a crystal ball. You need a feedback loop.

A simple way to keep yourself honest is to create a range:

$$\text{Enough (floor)} \leq \text{Enough (comfortable)} \leq \text{Enough (dreamy)}$$

If you can fund the floor, you have safety. If you can fund comfortable, you have a life you can enjoy. Dreamy is a bonus, not a requirement.

Actionable takeaway

Tonight, open a notes app or a blank page and write two numbers:

1. Your *monthly survival* estimate (housing + food + utilities + transportation + insurance/healthcare + essentials).
2. Your *monthly comfortable* estimate (survival + the specific extras you'd be sad to lose).

Don't research. Don't optimize. Guess with sincerity. Then add a third line: *What would I be willing to change—housing, work, location, lifestyle—to make the survival number unshakable?*

That question is where enough stops being abstract and starts becoming a plan you can actually live with.

1.3 The Truth About Your Net Worth

Net worth has a branding problem.

People hear the phrase and picture a glossy magazine profile: some billionaire's yacht, some founder's "estimated fortune," some celebrity's mansion doing all the talking. That's not what net worth is for. In your life, net worth is not a popularity contest and it's not a personality test. It's a diagnostic. It's the financial equivalent of stepping on a scale at the doctor's office: the number might not thrill you, but pretending it doesn't exist doesn't make you healthier.

If you started late, net worth can feel like a rude mirror. It shows the years you were paying bills instead of building assets, the time you were just trying to stay afloat. It can also show something else that's easy to miss when you're busy feeling behind: how much power you actually have. A clear baseline turns "I'm nervous" into "Here's what I'm working with."

The simplest definition is still the best:

$$\text{Net Worth} = \text{Assets} - \text{Liabilities}$$

Assets are what you own that has value. Liabilities are what you owe. The difference is your starting point.

Not your destiny. Your starting point.

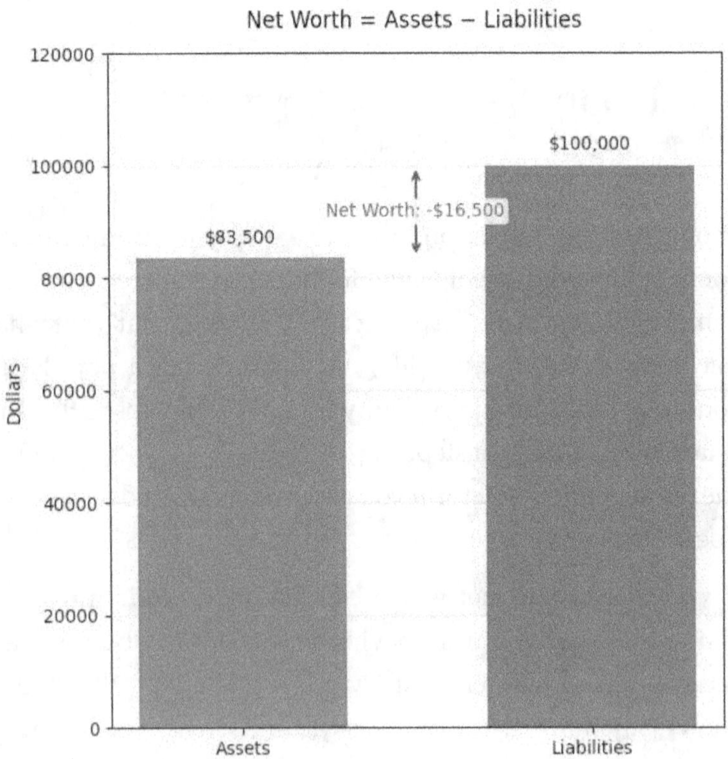

That little equation looks harmless until you actually do it. Then it becomes electric. Because now you have something most people never build: a personal balance sheet.

A balance sheet sounds corporate, but it's just a snapshot taken on a specific date. Think of it like a freeze-frame. It doesn't explain everything about your life, but it tells the truth about your financial structure at that moment. You can improve a structure. You can't improve a fog.

Why late starters need a balance sheet (and why it shouldn't hurt your feelings)

If you've ever avoided checking your accounts because you "already know it's bad," you're not alone. Avoidance is a normal response to discomfort. The problem is that avoidance turns a manageable situation into a mysterious monster. A monster is always scarier than a number.

Late starters benefit more than anyone from reducing mystery. When you have fewer working years left, you want fewer financial surprises. A balance sheet is how you stop being ambushed.

Also: net worth is not a moral scorecard. It's not a measure of effort, love, intelligence, or resilience. A caregiver can have a lower net worth than a single person with no responsibilities; that doesn't mean the caregiver "failed." It means resources went to real life. The balance sheet simply helps you see what choices are available now.

How to build your personal balance sheet in one sitting

Pick a date. Literally write it at the top: *Balance Sheet as of* __________. This matters because it frames the result as a moment in time, not a permanent label.

Then split a page into two columns:

- *Assets (what you own)*
- *Liabilities (what you owe)*

You don't need perfection. You need completeness at a level that's useful.

A template you can copy:

Assets	*Amount*
Checking / savings (total cash)	
Retirement accounts (401(k), 403(b), 457, IRA, Roth IRA)	
Brokerage / investment accounts (non-retirement)	
HSA balance (if invested, still your asset)	
Home value (estimate)	
Car(s) resale value (estimate)	
Other valuable property (optional: tools, equipment)	
Total Assets	

Liabilities	*Amount*
Credit cards (current balances)	
Student loans	
Auto loans	
Mortgage balance	
Personal loans / "buy now, pay later" balances	
Medical debt / payment plans	
Taxes owed (if applicable)	
Total Liabilities	

Net Worth (Assets - Liabilities)	

A few ground rules keep this from turning into a spiral:

- Use current balances, not "original loan amounts."
- For home value and car value, estimate. Pick a reasonable number and move on.

- Don't overcomplicate with pennies. Round to the nearest $100 or $1,000 if that keeps you moving.

The point isn't to win an accounting award. The point is to see the structure clearly enough to make better decisions.

What counts as an asset? (The ``can I sell it or use it later?" test)

Assets are things with real value that you could convert into money or use to fund your future. Cash is obvious. Retirement accounts are obvious, too—but they come with rules. Home equity can be real, but it's not a checking account. A car is an asset, but it's a melting ice cube: it tends to lose value.

A quick test: if you needed money, could you reasonably turn this into money *without* ruining your life? That helps separate meaningful assets from clutter.

That also brings up a subtle but important point: not all assets are equally *usable*. A dollar in a checking account can be spent today. A dollar in a traditional 401(k) is yours, but it's wrapped in tax rules and age rules. You still count it as an asset—it's absolutely part of your wealth—but when you later convert net worth into a retirement income plan, you'll treat different buckets differently.

What counts as a liability? (Yes, include the boring ones)

Liabilities are anything you owe: balances that have a claim on your future income. The obvious ones are credit cards, student loans, auto loans, and a mortgage. The sneaky ones are the payment plans and the "it's basically free money" deals that still require monthly payments.

Include all of them. Late starters can't afford to be surprised by a *small* liability that becomes a big drag when you're trying to ramp up savings.

A reality check that helps: a liability isn't only a number. It's a monthly payment attached to that number. And monthly payments are what squeeze your ability to invest.

A concrete case: Andre's negative net worth that wasn't a catastrophe

Andre is 44, works in IT, and earns $98,000. He feels like he's "doing okay" because he pays his bills and rarely carries credit card debt. Then he builds a balance sheet.

Assets:

- $7,500 cash
- $58,000 in a 401(k)
- $12,000 in a Roth IRA
- $6,000 car resale value

Total assets: about $83,500.

Liabilities:

- $0 credit cards
- $38,000 student loans
- $62,000 auto loan (new car, long term)

Total liabilities: $100,000.

Net worth: –$16,500.

Andre's first reaction is embarrassment. His second reaction is anger: "I make almost six figures and I'm still negative?" Then, because he has the numbers, he sees the actual story: the car.

That's the moment the balance sheet becomes useful. It doesn't scold him. It points.

He realizes the auto loan isn't just "a debt," it's a $980 monthly payment that eats the exact cash flow he needs to increase retirement contributions. He decides not to punish himself, but to restructure his plan: he sells the car, buys a reliable used one, and cuts his payment in half. The new balance sheet looks better, yes, but the real win is that his monthly breathing room increases. He redirects the freed-up cash into retirement contributions automatically.

His net worth didn't magically turn positive overnight. But his *trajectory* changed immediately. That's what the balance sheet is for: not to flatter you, but to give you leverage.

The ``net worth traps" that trip up late starters

Trap 1: Counting home value like it's cash. Home equity is real wealth, and it can matter a lot. But you can't buy groceries with a Zestimate. If your plan assumes you'll tap home equity later (downsizing, selling, a reverse mortgage, renting a room), that can be a valid strategy. Just don't confuse "wealth on paper" with "money available."

Trap 2: Ignoring the liabilities you hate thinking about. If you omit a debt because it makes you uncomfortable, you're not protecting yourself. You're protecting the debt. Debts thrive in the dark.

Trap 3: Treating retirement accounts as if they're all the same. A traditional 401(k) and a Roth IRA can both be assets, but they're not interchangeable. Traditional accounts are generally taxed when you withdraw. Roth accounts are generally funded with after-tax dollars and can have different tax treatment later. None of this needs to be mastered today, but it's worth respecting: part of "truth" is acknowledging constraints.

Trap 4: Believing a negative net worth is rare. It's not. Student loans, mortgages, and medical events can do that. A negative net worth is a signal, not a verdict. The worst thing you can do is interpret it as a reason to stop.

The most important number hiding inside net worth

Net worth is a snapshot. But it contains clues about the movie you're living in.

When you look at your assets and liabilities, ask one question that cuts through noise:

> Which line item has the most control over my future monthly cash flow?

Sometimes it's credit card debt because of the interest rate. Sometimes it's an auto loan because of the payment. Sometimes it's the mortgage because housing dominates the budget. Sometimes it's not even a debt—it's the fact that you have almost no liquid savings, which means every surprise becomes a mini-crisis.

That question turns net worth from "a number" into "a plan." It tells you where a change will have the biggest effect.

How often to update it (and what ``progress" really looks like)

Do the balance sheet once now, then update it quarterly for the first year. After that, twice a year is fine for most people. You're not trying to micromanage your life. You're building awareness.

Progress isn't only "net worth went up." Sometimes progress looks like:

- the same net worth, but less high-interest debt (a quieter financial life)
- the same net worth, but more cash reserves (less fragility)
- a higher net worth because liabilities dropped (often more meaningful than a lucky market year)

If your net worth rises because your investments grew, that's great. If it rises because you paid off a toxic debt, that's also great—and often more within your control.

Actionable takeaway

Within the next hour, build your first draft balance sheet on one page. Then circle *one* line item—asset or liability—that, if improved over the next 90 days, would make you feel noticeably less financially cornered.

Write a single sentence under it: *"If I move this number, my life gets easier because ________________________________.*" That sentence is the beginning of momentum, and momentum is what late starters can't afford to waste.

1.4 Cash Flow: The Pulse of Your Plan

Net worth tells the truth about your financial *position*. Cash flow tells the truth about your financial *power*.

You can have a respectable net worth and still feel trapped, because your monthly obligations leave no room to breathe. You can also have a modest net worth and be quietly dangerous—in the best way—because you've built a surplus that shows up every single month like clockwork. For late starters, that surplus is the raw material that becomes retirement contributions, debt payoff, and options. Without cash flow, every plan is just a wish with nicer formatting.

Cash flow is simply the movement of money in and out over time. Income comes in. Expenses go out. The difference is your surplus (if positive) or deficit (if negative). That's it. No jargon. No mysticism. The only reason it feels hard is that real life is messy: income is irregular, expenses don't arrive politely, and the brain prefers not to count.

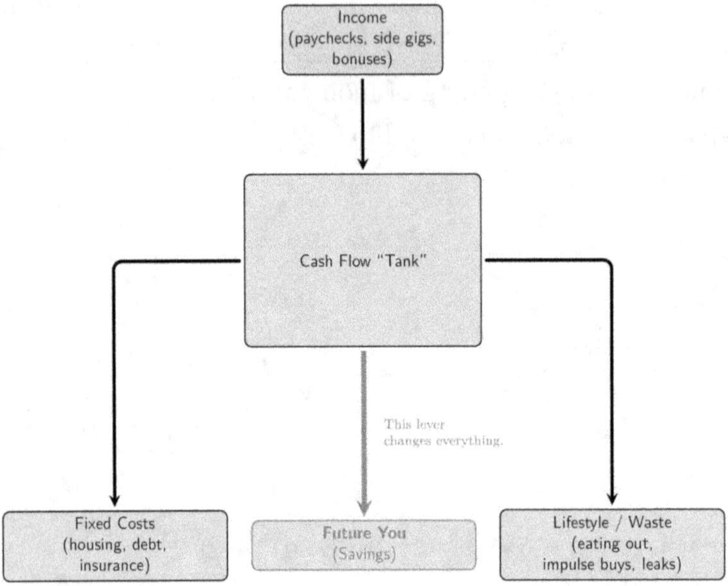

That little diagram is the whole game. Not because it's cute, but because it forces a question most people avoid: *Which pipe is getting the best of me?* If your fixed-cost pipe is huge, you're living on a narrow ledge. If your lifestyle/waste pipe is quietly massive, you're funding forgettable pleasures with unforgettable consequences. And if your Future You pipe is a drip, it doesn't matter how much you *believe in retirement*. Belief doesn't compound. Contributions do.

1.5 The late starter's advantage lives in cash flow

The earlier sections did something important: they got you to face the emotional reality of starting late, and they gave "enough" a definition rooted in your life instead of the internet's. That work matters because it turns cash flow from punishment into purpose. When you know what you're trying to fund—safety, dignity, and the ability to choose—redirecting money stops feeling like deprivation and starts feeling like buying options.

Late starters also tend to have a hidden asset: income at its peak or near-peak. Maybe you're finally earning what you *should* have been earning years ago. Maybe you're in a dual-income household for the first time. Maybe childcare costs are about to drop. Those transitions show up as cash flow opportunities if you catch them. If you miss them, they disappear into lifestyle creep: the upgraded car, the nicer neighborhood, the "we deserve it" spending that feels harmless because it's incremental.

Cash flow is where you intercept those moments.

1.6 Why budgets fail (and how to do one that won't)

A budget fails when it's built like a fantasy novel: heroic intentions, no plot realism.

Most budgets are either too strict to survive a normal week, or too vague to change anything. They also tend to be built backward: people decide what they *want* to spend, then hope reality cooperates. Cash flow planning works better when you treat the past as evidence.

A workable process has three steps:

1. *Collect the facts.* Use the last 30–90 days of transactions, plus your most recent pay stubs.
2. *Name the categories that matter.* Not 27 categories. The few that actually drive your life.
3. *Compute the gap.* Income minus expenses. Then decide, intentionally, where the gap goes.

Notice the order: facts first, then choices. Late starters don't need motivational posters. They need clarity and a system that can run even when life gets busy.

1.7 Income: count it the way your life receives it

Start with your net income—what lands in your bank account after taxes, insurance, and payroll deductions. That's the money you can actually direct.

If your income is regular, great: use a typical month. If your income is irregular (commissions, bonuses, contract work), don't torture yourself with optimism. A practical method is to take last year's total take-home pay and divide by 12 to estimate a monthly average. Then set aside a buffer in strong months so weak months don't force you into credit card debt. This is less exciting than assuming every month will be a good month, and infinitely more effective.

Also: don't ignore "once-a-year money." Tax refunds, bonuses, and RSUs have a habit of vanishing because they don't feel like income. For late starters, those chunks can become your plan's turbo button *if* you decide in advance where they go.

1.8 Expenses: stop arguing about morality and start sorting by behavior

Expenses fall into three buckets that matter for control:

- *Fixed costs*: hard to change quickly (housing, minimum debt payments, insurance).
- *Variable costs*: can be adjusted with attention (food, gas, utilities).
- *Discretionary costs*: the stuff that expands to fill your stress, your boredom, and your phone battery (shopping, dining out, convenience).

This isn't about judging purchases. It's about identifying what responds to pressure.

Late starters get the biggest payoff from the expenses that are (a) recurring and (b) easy to autopilot. A $12 subscription you forgot for two years is a perfect example: it's not huge, but it's pure leakage. A car payment is another: it's huge and recurring, which means changing it can free up hundreds per month for Future You.

1.9 A concrete case: Sam's ``I make good money, where is it going?" moment

Sam is 41, single, and earns $124,000 as a mid-level manager. He has $52,000 in a 401(k), $14,000 in cash, and no credit card debt. He also feels behind and can't understand why he can't "seem to save."

He finally does the unglamorous work: he prints two months of bank and credit card statements and highlights every transaction. The pattern is almost funny in hindsight.

Fixed costs:

- Rent: $2,450
- Car payment + insurance: $860
- Student loan: $410
- Utilities + phone + internet: $260

Variable:

- Groceries: $520
- Gas: $160

Discretionary:

- Restaurants and delivery: $1,150
- "Small" online purchases: $640
- Weekend spending (bars, events, rideshares): $580
- Subscriptions: $140

Sam isn't irresponsible. He's exhausted. His spending is the shape of a busy life where convenience is the default reward. The key is that the money isn't going to something he values deeply. It's going to relief.

So he makes a plan that respects psychology, not just arithmetic:

- He sets a weekly "fun money" transfer to a separate debit card. When it's gone, it's gone. No drama. No guilt.

- He caps delivery to one night per week and stocks two "emergency meals" in the freezer for late nights.
- He increases his 401(k) contribution by enough to reduce take-home pay by $500/month, because he knows if the money hits checking, it will evaporate.

He doesn't become a monk. He becomes intentional. Three months later, he has a consistent surplus and his anxiety drops because the plan is no longer based on hope.

1.10 The most important calculation: your monthly surplus

Once you have income and expenses, do the simplest math in personal finance:

$$\text{Monthly Surplus} = \text{Monthly Income} - \text{Monthly Spending}$$

If that number is negative, the goal is not to "invest." The goal is to stop bleeding. You can't out-invest a cash flow deficit. If the number is positive, you now have the fuel that turns "enough" into something real.

Late starters should treat the surplus like a job offer: you are hiring your money to work for your future. If you don't assign it, it will assign itself. Usually to lifestyle.

1.11 How to route the surplus (the order that keeps you calm)

Here's an order that tends to work for late starters because it reduces fragility first:

1. *Stabilize the present:* build a small emergency buffer (even $1,000–$2,000) if you don't have one. This prevents every surprise from becoming a credit card event.
2. *Kill toxic debt:* pay down high-interest debt aggressively. Interest is a guaranteed negative return.
3. *Fund Future You:* increase retirement contributions automatically. Make it hard to undo.

If you already have an emergency fund and no high-interest debt, you can push more directly into retirement. The point is to route money in a way that reduces stress, because stress is what causes backsliding.

1.12 Automation: the late starter's best friend

If willpower were reliable, nobody would be behind.

Automation is simply deciding once and letting the system repeat the decision. You can automate:

- retirement contributions through payroll
- transfers to savings accounts
- extra payments toward a specific debt

Payroll automation is particularly powerful because it prevents the "I'll move it later" problem. The money never becomes available for lifestyle spending. It becomes invisible, which is exactly what you want for long-term goals.

A useful mindset: treat savings like a bill. Not because you're depriving yourself, but because you're paying for a future life you want to live.

1.13 A reasonableness check (without comparing your life to strangers)

Sometimes people do a cash flow audit and feel like their spending is "abnormally high." Before you spiral, remember: spending is shaped by location, family size, health, and a dozen invisible constraints. Benchmarks can be helpful as a reasonableness check, but your life is not a national average.

The point of the audit is not to become "normal." It's to identify what you can control and what you can't, then design a plan around the truth. If childcare is crushing your budget for the next three years, the plan might be to tread water temporarily while protecting retirement contributions at a minimum level. That can still be success if it's intentional. Cash flow planning is not about perfection. It's about steering.

1.14 The ``Future You" pipe is a lever, not a leftover

Most people treat savings as what happens if there's money left. Late starters can't afford that approach. Savings has to be a lever you pull on purpose.

That doesn't mean you save an extreme amount immediately. It means you build an escalation plan: increase contributions in steps that your life can absorb. A common method is to tie increases to raises: each time your income rises, you "tax" the raise for Future You before lifestyle expands to swallow it.

If you get a 4% raise, you might send 2% to retirement and keep 2% for today. This is how you build intensity without misery.

1.15 Actionable takeaway

In the next 72 hours, run a *two-month cash flow audit*:

1. Add up your net income for the last two months.
2. Add up your total spending for the last two months (use statements; don't guess).
3. Compute the difference and divide by two to get your average monthly surplus (or deficit).

Then do one decisive move with that result: if you have a surplus, increase your automatic "Future You" transfer by *25% of that surplus* starting next payday. If you have a deficit, pick the single biggest recurring expense you can change within 30 days and reduce it enough to break even.

A plan you can repeat is worth more than a plan you admire. When your cash flow starts producing a reliable surplus, retirement stops being a scary headline number and becomes a monthly habit you can actually feel.

1.16 Rewiring Your Money Mindset

A strange thing happens when you finally decide to take retirement seriously in midlife: the math is rarely the hard part.

Yes, there are real constraints. Yes, you may have fewer years and more responsibilities. But the reason people don't act is usually not because they can't add or subtract. It's because money triggers ancient circuitry: fear, avoidance, status, shame, and the particular dread of making the "wrong" move when you feel you don't have time to recover. Your brain treats financial uncertainty the way it treats a threat in the dark. It wants you to freeze, conserve energy, and wait for more information. Unfortunately, retirement doesn't give extra credit for waiting.

The good news is that mindset isn't a motivational poster. It's wiring. Wiring can be rerouted with the right levers: reducing decision load, shaping defaults, and making the future feel real enough that you stop sacrificing it to today's discomfort.

The ostrich instinct: why smart adults avoid the numbers

If you've ever delayed opening a retirement statement because you "weren't in the mood," you're not irrational. You're human.

Avoidance is protective in the short term. When life is already full—work deadlines, aging parents, kids, health stuff, relationship stuff—your brain doesn't want one more problem. It wants to preserve bandwidth. Bandwidth is the mental space you use for planning, focus, and self-control. When money stress rises, bandwidth shrinks. You become more reactive, less patient, and less able to do the slow, boring tasks that improve your future.

This is why late starters often live in a loop: fear triggers avoidance; avoidance allows the problem to grow; the larger problem creates more fear. It's not laziness. It's a predictable stress response.

The antidote isn't to "be more disciplined." The antidote is to make the first actions so small, so automatic, and so emotionally safe that your brain doesn't resist.

One practical rule: *never begin with the most painful number.* Begin with something neutral and concrete—like logging into your payroll system to see your current 401(k) contribution rate. That's a fact, not a judgment. Once you're in motion, you can handle more.

Loss aversion: why change feels worse than staying stuck

There's a psychological quirk that hits especially hard after 40: loss aversion. It means losses feel more painful than equivalent gains feel good. Losing $100 hurts more than gaining $100 delights. That's not a moral flaw; it's built in.

Loss aversion explains why people keep expensive habits they don't love. Cutting them feels like losing. Even if the trade is objectively favorable—less spending now for more freedom later—your brain experiences it as a threat to your quality of life. The trick is to stop framing changes as deprivation and start framing them as purchases.

You are not "giving up dining out." You are buying the ability to retire without panic. You are not "driving an older car." You are buying breathing room. You are not "being cheap." You are buying options.

The reframe only works if it's specific. "Freedom" is too vague. "A paid-off car and $800 more per month into retirement" is a concrete purchase.

A useful exercise: when you decide to cut an expense, write down what you are buying with that cut. Not in general. In dollars.

I am cutting $150/month to buy $150/month of Future Me.

This may sound simplistic, but it changes the emotional valence. Instead of a loss, it becomes a trade.

The paralysis trap: ``I'm afraid to do the wrong thing"

Late starters often get stuck on a deceptively noble idea: *I want to make the best decision.* That's how people end up reading articles for six months and contributing nothing.

Perfectionism is often fear wearing a suit. When time feels short, the cost of mistakes feels high, so you delay until you're "sure." But the market doesn't pay you for certainty. It pays you for time in the game.

Here's a sharper way to think: the worst financial decision for most late starters is not choosing a slightly suboptimal fund. It's choosing *inaction*. If you invest in a reasonable, diversified way and keep contributing, you can correct course later. If you wait, you can't buy back those months.

A mindset shift that helps: *Choose a decision that is easy to reverse.* Increase your payroll contribution by 1%. You can lower it if you must. Open an IRA with a target-date fund. You can change investments later. The goal is to get the machine running.

Mental accounting: using your brain's quirks instead of fighting them

People like to believe money is money, and every dollar is identical. In practice, we label dollars. Rent money feels different than vacation money. A tax refund feels like "bonus money." Cash in checking feels spendable; cash in a separate account feels protected. This is called mental accounting: our brains create buckets.

You can roll your eyes at that, or you can use it.

Late starters benefit from deliberate buckets because buckets reduce decision fatigue. When everything sits in one checking account, every purchase becomes a negotiation: *Can I afford this? Should I buy this? What about retirement?* Negotiations are exhausting, and exhaustion is where plans go to die.

A simple bucket system that works:

- *Bills Account*: fixed costs auto-paid

- *Spending Account*: guilt-free day-to-day spending
- *Future You Account*: retirement contributions and/or a separate savings transfer

You are not doing this because you can't be trusted. You're doing it because your brain is a finite resource. The fewer times you force it to make a hard choice, the more consistent you become.

Future-you continuity: why your 67-year-old self feels like a stranger

One reason saving is hard is that the benefits go to someone you can't quite feel. Your future self is real, but emotionally, they can feel like a stranger with your name. When the future self is blurry, the present self wins every argument. "I'll start next month" feels painless. "I'll cut spending later" feels reasonable. Then later arrives with the same habits and fewer options.

The goal is not to scare yourself. The goal is to make the future self familiar.

A surprisingly effective technique is to write a short, specific description of a day in your retirement. Not fantasy travel. A normal Tuesday. Where do you wake up? What do you eat? What do you do with your morning? What annoys you? What comforts you? The more ordinary and sensory it is, the more your brain treats that person as you.

Then connect money to that day in one sentence: *"I'm investing $400 per month so that Tuesday can be calm."*

This is not woo. It's practical psychology. The future becomes easier to protect when it feels real.

A concrete case: Leah's ``financial freeze" and the two-minute rule

Leah is 47, a teacher, and one of the most competent people in her building. She runs committees, mentors new staff, and manages classrooms with the calm authority of someone who has seen everything. Money, however, makes her feel fourteen.

Her husband handles most bills. After a tough year with medical expenses, Leah starts waking up at 3 a.m. with the thought: *What if we can't retire?* In the daylight she avoids it. She's "too busy." She'll "deal with it this summer." Summer arrives and vanishes.

Leah doesn't need more information. She needs a different on-ramp.

She adopts a rule: *two minutes counts.* The goal is not to finish the whole task. The goal is to touch the system daily so avoidance loses its power.

Day 1: she finds her latest retirement statement in her email. She doesn't open it. She just finds it. Day 2: she opens it and writes down the balance. That's it. Day 3: she logs into payroll and checks her contribution percentage. Day 4: she increases it by 1%. Day 5: she tells her husband they're going to do a net worth snapshot on Saturday, with coffee, no blame, and a timer.

Within two weeks, Leah feels a shift that surprises her: the dread shrinks. The problem didn't disappear, but it became finite. She moved from "something is wrong" to "this is the number, and we're acting."

That's the power of tiny actions. They don't solve everything. They break the spell.

The ``late-starter" emotional cocktail: shame, urgency, and status

After 40, money decisions are rarely isolated. They sit next to identity. You may be managing a career, being a parent, being a partner, being the dependable one. Discovering you're behind financially can feel like discovering you've been failing at adulthood. Shame rushes in, and shame is a terrible planner.

Shame also tends to produce performative decisions: buying status to soothe insecurity. The upgraded car "because I deserve something." The expensive vacation "because we're stressed." The home renovation "because everyone else's looks better." These aren't bad choices in themselves. The problem is when they're made unconsciously, as emotional anesthesia.

A powerful line to keep in your pocket:

> *Spending is not self-care if it steals peace from your future.*

Real self-care in midlife often looks boring: building an emergency fund, automating contributions, paying down a debt, saying no to a lifestyle upgrade that would keep you working longer than you want.

How to build a system that doesn't require you to be ``good at money"

The most important mindset shift for late starters is this: stop expecting motivation to carry you. Motivation is weather. Systems are infrastructure.

A system has three qualities:

- *Automatic*: it runs without repeated decisions
- *Visible*: you can tell if it's working quickly
- *Forgiving*: one bad month doesn't collapse it

Automation was already discussed as a cash flow tool. Here's the psychological reason it matters: each time you decide whether to save, you invite an internal debate. Debates are where fear wins. Automation removes the debate.

Visibility matters because progress reduces anxiety. A simple graph of your account balance, updated monthly, can do more for your consistency than any motivational speech. You're not worshipping the number; you're watching the machine.

Forgiveness matters because midlife is unpredictable. A system that requires perfect adherence will fail the moment life happens. Build one that can handle a bad month without turning into "welp, I blew it."

The decision diet: fewer choices, better outcomes

If you want to feel calmer about money, cut the number of financial decisions you make in a week.

Some examples of decisions you can eliminate:

- *checking your accounts daily* (it spikes anxiety without improving outcomes)
- *choosing investments constantly* (it encourages tinkering)
- *deciding every night whether to cook or order* (make a default)

Defaults are not boring; they are sanity. When you reduce decisions, you reduce the opportunities for stress to hijack your plan.

A practical approach is to set "money office hours." One hour per week, same time, same place. During that hour you pay bills, check progress, and make any adjustments. Outside that hour, you don't ruminate. You live your life. The boundary itself reduces background anxiety.

Actionable takeaway

Pick one psychological trap that matches your pattern and install a countermeasure today:

- If you avoid the numbers, use the *two-minute rule*: spend two minutes touching your finances (log in, write one balance, file one document). Two minutes counts.
- If you fear making the wrong decision, choose one *easy-to-reverse* action: increase your retirement contribution by 1% for the next two paychecks.
- If you leak money through stress spending, create a *spending bucket*: transfer a fixed weekly amount to a separate card/account and spend from there without guilt.

Then write one sentence and stick it somewhere you'll see it: *"My plan doesn't need me to feel ready. It needs me to act consistently."* Consistency is how you turn fear into a tool: it becomes the signal that you should do the next small action, not the signal that you should hide.

Chapter 2

Creating Margin: Cash Flow, Debt, and Income

At 40, we don't have time for financial guilt trips or complex spreadsheets that require a PhD to maintain. We need margin-that vital breathing room between what comes in and what goes out. This chapter isn't about giving up lattes to save pennies; it's about building a robust cash-flow engine that automatically funds your catch-up plan while efficiently crushing the debt that keeps dragging you backward.

2.1 Budgeting for the Time-Crunched

If your calendar looks like a game of Tetris and your brain feels like it has *exactly one* remaining tab open, traditional budgeting advice can sound almost insulting. Track every purchase. Categorize every receipt. Reconcile weekly. Cook from scratch. Also meditate, stretch, and get eight hours.

Most people over 40 don't fail at budgeting because they're irresponsible. They fail because the system they were handed assumes time is free and attention is unlimited. It's neither. You don't need a budget that proves you were "good." You need a spending plan that quietly makes the right thing happen while you live your life.

The trick is to stop treating budgeting like an accounting project and start treating it like plumbing. You don't stand in your kitchen every day thinking about pipe diameters. You set up the pipes so water goes where it should. A good budget is the same: income comes in, the important things get fed first, and the rest is allowed to be spent without guilt—because the plan already did its job.

Why tracking breaks the exact people who need it most

Detailed tracking has a hidden requirement: a stable life. If your expenses are predictable, your schedule is predictable, and your emotional bandwidth is predictable, then yes, you can track every dollar and feel in control.

But midlife tends to be the opposite. The kid needs braces, the car is aging, your parents might be leaning on you, your job might be changing, and your body starts sending invoices you didn't authorize. The result is what I'll call *budget fatigue*: the plan is technically correct, but maintaining it becomes a second job. Once it slips for a week, it feels ruined. Once it feels ruined, you avoid looking. Avoidance turns into drift. Drift turns into debt.

There's also a more personal problem: hyper-tracking makes perfectly normal spending feel like moral failure. The latte becomes a character flaw. The birthday gift becomes "why can't you be disciplined?" That shame is not motivating; it's corrosive. And it trains you to associate your finances with self-criticism instead of problem-solving.

The goal isn't to become a better tracker. The goal is to create *margin*—room to breathe in your monthly cash flow—so your future isn't decided by a random Tuesday.

Reverse budgeting: pay yourself first, then live your life

Reverse budgeting flips the usual script. Instead of monitoring everything you spend and hoping there's money left over for savings, you *fund the priorities first*. What remains is yours to spend—no spreadsheets required.

Think of it as a bouncer at the door of your paycheck. Before money enters your lifestyle, it has to show ID and prove it's going to the right places.

Here's the core structure that works for time-crunched people because it's simple, durable, and hard to "mess up":

1. *Fixed Needs*: bills that keep life functioning (housing, utilities, basic groceries, transportation, insurance, minimum debt payments).
2. *Future You*: the money that buys freedom later (retirement contributions, emergency savings, sinking funds, extra debt payoff).
3. *Guilt-Free Spending*: everything else (restaurants, subscriptions, nicer groceries, hobbies, convenience, travel, the stuff that makes adulthood enjoyable).

Notice what's missing: a demand that you perfectly categorize 37 subtypes of "household supplies." Reverse budgeting doesn't care whether the Target run was $62 of toiletries and

$18 of impulse candles. It only cares that Target didn't steal money from your future.

This approach also fixes a common psychological trap: *the monthly reset illusion*. People treat the first day of the month like a magical eraser. If you blew the budget on the 7th, you'll "be good" for the rest of the month. Except life doesn't stop. Reverse budgeting doesn't rely on a fresh start. It relies on order of operations.

The three-bucket plan you can run in under 20 minutes a week

The point isn't to make this pretty. The point is to make it run without you.

Start with your net monthly income (what actually hits your checking account). If you're paid biweekly, multiply one paycheck by 26 and divide by 12 to estimate a monthly number. You can refine later; perfection is not required to begin.

Then do a fast sort:

- List your *Fixed Needs* as a single monthly number. If you don't know it, don't panic—pull the last two months of bank statements and count only the essentials. Round up. You're building a safe container, not a courtroom case.
- Decide on a *Future You* number. This is the heart of reverse budgeting: it's a commitment, not a leftover.
- Whatever remains is *Guilt-Free Spending*.

If you want a starting template, a lot of people recognize the rough logic of "50/30/20." For time-crunched catch-up savers, it often needs a twist: the "Future You" bucket is the engine, and the guilt-free bucket is the stabilizer. You don't starve the stabilizer; you right-size it.

A practical starting point for many households is:

- Fixed Needs: 55–70% (depending on housing and insurance costs)
- Future You: 15–25% (higher if you're closing a retirement gap)
- Guilt-Free Spending: the remainder

If your Fixed Needs are currently swallowing 80% or more, you're not broken. You're just operating with a thin runway. That doesn't mean reverse budgeting can't work; it means your first wins may come from small structural changes (insurance shopping, subscription cleanup, refinancing debt, or income increases) rather than from "trying harder."

A real-world example: Maya and Chris, two kids, one surprise after another

Maya (45) and Chris (47) look like they have it together. Two kids, both working, decent incomes. But every month felt like a magic trick: money disappeared, and they couldn't tell you where it went. They tried tracking apps. Maya kept up for ten days; Chris forgot on day two. Then the guilt set in, and they avoided their accounts until the next paycheck arrived like a rescue helicopter.

Their numbers:

- Net income: $8,600/month

- Fixed Needs (mortgage, utilities, insurance, car payments, basic groceries, childcare, minimum debt payments): about $6,200/month
- Current saving: "whatever is left," which averaged $200/month but often got wiped out by surprises

The problem wasn't a lack of income; it was that their lifestyle had first access to their cash flow. Savings was the guest who only got fed if everyone else was full.

They set a *Future You* target of $900/month. Not because it felt easy—it didn't—but because it was the smallest number that would change their trajectory.

Here's what made it workable:

- $500/month went automatically to retirement (split across workplace plan and Roth IRA contributions, depending on what was available).
- $250/month went into a separate savings account labeled "Car & House."
- $150/month went into an "Oh no" fund (their starter emergency fund).

Now the plan had gravity. They didn't need to debate savings every month. It happened before opinions formed.

What changed *immediately* wasn't their spending; it was their stress. The questions at the grocery store stopped being "Can we afford this?" and became "Is this coming out of guilt-free spending or fixed needs?" That's a smaller question, and it's one you can answer while holding a cart and a tired child.

The second month, the minivan needed $780 in repairs. Previously, that would have landed on a credit card and become a long, expensive tail. Instead, the "Car & House" account covered it. They still felt annoyed (because car repairs are a tax on having a life), but they didn't feel trapped. That distinction matters. Annoyed is survivable. Trapped is how people quit.

The only tracking that really matters: one weekly check

Reverse budgeting is not "never look at your spending." It's *look in the highest-leverage place*. For most people, that place is not a category report; it's the total available in the guilt-free bucket.

Here's a simple weekly rhythm that takes five minutes:

1. Open your checking account.
2. Look at what's left after upcoming bills.
3. Decide whether the next seven days need a "normal" mode or a "tighten up" mode.

This works because it matches how life is lived. You don't spend by category; you spend by week. Weekends have different gravity than Wednesdays. School weeks differ from travel weeks. A weekly check respects reality.

If you want one extra layer of clarity, create a dedicated spending account (or a second checking account) for guilt-free spending and transfer a fixed amount into it each payday. Then your "tracking" becomes glancing at one number. When it's low, you naturally slow down. When it's healthy, you spend without the background hum of financial anxiety.

This is why reverse budgeting is friendly to real humans: it makes the right choice the default choice.

Fixed needs: where time-crunched people accidentally overpay

Fixed costs feel like granite, but a surprising amount of granite is actually foam painted gray. When you're busy, you accept renewals and default settings. Companies love default settings.

You don't need a heroic frugality phase. You need one or two high-impact moves per year:

- *Insurance*: premiums creep up quietly. Shopping once a year can produce meaningful savings without changing your life.
- *Housing*: if refinancing or negotiating property taxes/assessments is feasible, it can create monthly margin bigger than any grocery optimization.
- *Subscriptions*: not because $9.99 is killing you, but because six $9.99 charges become $60 of invisible friction.
- *Phone and internet*: loyalty is often punished. A 30-minute call can be worth more than hours of coupon clipping.

There's a reason this matters for the over-40 crowd: time has a higher opportunity cost now. The best savings aren't the ones that require constant attention; they're the ones that reduce the baseline permanently.

A good rule: if a change saves less than $20/month but requires ongoing effort, it's probably not worth your limited energy. If a change saves $100/month and you can set it once and forget it, that's the sweet spot.

Future You: make it a bill, not a wish

"Pay yourself first" is an old phrase, but most people treat it like a motivational poster. The real power comes when you treat Future You like a non-negotiable bill—because it is. It's the bill for your own independence.

If retirement saving feels abstract, translate it into something concrete: Future You is the person who wants choices. Choices about work. Choices about health. Choices about where to live. Without savings, the choices shrink and the stress grows.

Time-crunched people often wait for a month that feels "calm" to start saving more. Calm months are rare. The better strategy is to set a number that's *slightly uncomfortable but doable* and then let your lifestyle adjust around it. Humans adjust faster than they think when the decision is made once instead of debated every day.

If you're behind and feel the urge to go big immediately, resist the boom-and-bust cycle. Pick a number you can maintain for a year. Consistency beats intensity, especially when the goal is to build a system you can run while life happens.

One practical move that helps: tie Future You contributions to payday, not month-end. Payday is when the money is real and available. Month-end is when the money has been negotiated away by everything else.

Guilt-free spending: the bucket that keeps you from rebelling

A plan that feels like punishment will be abandoned. That's not a character flaw; it's a design flaw.

Guilt-free spending is not the enemy of progress. It's the part that makes progress sustainable. It gives your present self a reason to cooperate with your future self.

The key is that guilt-free spending needs *boundaries*. Not moral boundaries—cash boundaries. If you want a latte, you can have it. If you want to eat out twice this week, go for it. The only rule is: it happens inside the amount you already decided. When the bucket is empty, you don't spiral into shame; you switch modes.

This is where reverse budgeting quietly heals the emotional relationship with money. You're not asking "Am I allowed?" You're asking "Is this worth it compared to everything else I could do with this bucket?" That's adulthood. That's agency.

If you live with a partner, guilt-free spending also reduces fights. When each person has a small personal allowance inside the plan, you stop litigating the tiny purchases that symbolize bigger fears. Money arguments are often "control arguments" in disguise. A clear guilt-free bucket gives both people room to breathe.

When income is irregular: build the plan on your worst month

If you're self-employed, commission-based, seasonal, or juggling side income, traditional budgets can feel like they were designed by someone who gets paid the same amount every two weeks and has never met a slow quarter.

Reverse budgeting still works, but you base it on a conservative number: your lowest reliable monthly income from the last 12 months (or the last six, if that's all you have). That number funds Fixed Needs and a modest Future You contribution. When you earn more, the surplus gets assigned deliberately: some to sinking funds, some to debt, some to taxes (if applicable), some to guilt-free spending.

This approach protects you from the most common irregular-income trap: spending like it's a good month right up until it isn't. A good month should feel like a tailwind, not a permission slip to increase your baseline.

A short, practical setup plan you can actually do this weekend

You don't need a financial makeover montage. You need a clean start with minimal steps.

1. *Pick your three numbers.* Fixed Needs (estimate), Future You (choose), Guilt-Free (what remains).
2. *Create one separation.* Either a second checking account for spending or a separate savings account for Future You. One separation changes behavior more than ten resolutions.
3. *Automate one transfer.* On payday, move the Future You amount out of your main spending stream.
4. *Schedule one weekly glance.* Same day, same time, five minutes. Treat it like brushing your teeth: not dramatic, just protective.

If you do only those four things, you will have something most people never get: a plan that works even when your week goes sideways.

You'll still have expensive months. You'll still have surprises. The difference is that the surprises won't automatically become debt, and the expensive months won't automatically become self-judgment.

Your finances don't need you to be perfect. They need you to be consistent in one place: paying Future You before the world gets the first bite.

2.2 Planning for the Unexpected

The moment you start running a cleaner spending plan, life tends to test it. Not in a dramatic, movie-plot way. In the quiet, annoying way: the dentist says "we should do this now," the car makes a new noise, the school announces a "suggested" donation that somehow feels mandatory, and your annual insurance premium arrives like it's delivering bad news with a smile.

Most financial stress isn't caused by true emergencies. It's caused by *irregular* expenses colliding with a monthly paycheck. The collision is what makes people reach for a credit card even when they're generally responsible. The problem isn't irresponsibility; the problem is physics. A once-a-year bill hitting a once-a-month budget creates a spike. Spikes create panic. Panic creates debt.

You can't stop the bills from existing. What you can stop is the surprise.

The difference between an emergency and an irregular expense

An emergency is something you didn't see coming and couldn't reasonably schedule: a medical event, a job loss, a broken furnace in the middle of winter.

An irregular expense is something you *know* will happen, but you don't know exactly when—or you know the month but not the mood you'll be in when it arrives. Car repairs. Home maintenance. Holiday spending. Back-to-school costs. Annual subscriptions. Professional dues. Gifts. Vet bills. Travel to weddings and funerals. Even routine healthcare costs can behave irregularly if your plan resets mid-year or your kid suddenly needs a specialist.

Irregular doesn't mean optional. It means the timing is annoying.

When irregular expenses aren't planned for, they get mislabeled as "emergencies." And once everything is an "emergency," the emergency fund gets drained, then rebuilt, then drained again. That cycle is exhausting. It also creates a subtle belief that saving "doesn't work," because you never get to keep the money.

The fix is to stop treating known irregulars like monsters under the bed. Turn on the light. Name them. Feed them on purpose.

Sinking funds: boring, brilliant, and weirdly calming

A sinking fund is a small pool of money you build slowly for a specific upcoming cost. You don't invest it; you don't gamble with it; you don't pretend it isn't yours. You drip money into it monthly, so that when the bill arrives you pay it like a normal person, not like a desperate one.

The name sounds like a ship problem, but the concept is the opposite: it keeps you afloat.

Here's the core idea in one sentence: *irregular expenses become manageable when you turn them into monthly expenses.*

If you know you'll spend $1,200 a year on car maintenance and repairs, your sinking fund target is $100 a month. If your annual auto insurance is $1,800, your target is $150 a month. The bill doesn't get smaller, but your stress does. And because the money is already sitting there, the credit card doesn't get invited to the party.

Sinking funds also protect your "Future You" savings. Without them, you end up raiding the accounts that were supposed to change your life later, just to survive the present. With them, the plan stops getting derailed by predictable chaos.

Why credit cards become the default (even for people who pay them off)

Credit cards are easy in the moment and expensive in the afterlife. The afterlife is the part where the balance rolls, interest starts compounding, and minimum payments quietly colonize your cash flow.

Irregular expenses are one of the main reasons people slide from "I pay my card off every month" into "I'm carrying a balance and I don't know how it happened." Nothing happened. The calendar happened.

A single $900 bill can be harmless if you have $900 set aside. It can be a financial infection if you don't. Once it's on the card, it competes with groceries and gas and the electric bill. It becomes sticky. And if another irregular expense hits before it's gone—and it will—you're suddenly in a debt loop that feels like it came out of nowhere.

Sinking funds are the quiet antidote. They don't require willpower in the moment of the bill. They require a small decision earlier, when you're calm.

The "known unknowns" list: what belongs in a sinking fund

Start by thinking in categories of life rather than budget categories. You're not building a spreadsheet. You're building a shock absorber.

Common sinking funds for a midlife household include:

- *Car & Transportation*: repairs, tires, registration, oil changes, eventual replacement down payment
- *Home*: maintenance, small repairs, appliance replacement, annual servicing (HVAC, gutters, pest control)
- *Health*: deductibles, prescriptions, dental, glasses/contacts, therapy co-pays
- *Gifts & Holidays*: birthdays, holidays, teacher gifts, charitable giving you want to do
- *Kids*: activities, sports fees, camps, school fundraisers, field trips, graduation costs
- *Annual/Quarterly Bills*: insurance premiums, professional dues, subscriptions you keep on purpose

- *Travel & Family Obligations*: weddings, funerals, family visits that are emotionally important but financially disruptive

You don't need ten funds. Most people do well with three to five. Too many becomes fussy and you stop maintaining it. The purpose is *predictability*, not perfection.

If you're not sure what's been knocking you off balance, scroll through the last 12 months of transactions and circle anything that made you say, "Oh, right." Those are your candidates. Your bank statements are basically a diary of your surprises.

How much should you put in each fund? Use math, then use mercy

The math is simple:

$$\text{Monthly Sinking Fund Contribution} = \frac{\text{Estimated Annual Cost}}{12}$$

If the annual cost is hard to estimate, use a range and start with the low end. You can adjust. Starting matters more than being correct.

Then use mercy: don't punish yourself with an impossible target. A sinking fund that's so large it squeezes your grocery budget will fail, and the failure will teach the wrong lesson. Better to start with "half-funding" and build up.

For example, if you estimate $2,400/year in car-related costs ($200/month) but you can only spare $100/month, start there. You're not pretending the other $100 doesn't exist; you're acknowledging reality while reducing damage. Half-funding still cuts credit card reliance in half *over time*, because some bills will land when you've built the fund up.

Also: sinking funds are not meant to cover catastrophe. That's the role of an emergency fund. Sinking funds are meant to cover the annoying, normal, expensive texture of adult life.

A case study: David, one roof, two parents, and the tyranny of ``annual"

David is 52, divorced, and the default adult for two aging parents. His finances aren't a mess, but they're constantly under pressure. He earns $6,200 net per month. His fixed bills are around $4,700. He contributes to retirement, but every few months something forces him to pause: a parent's plane ticket, a property tax bill, new tires, a dental crown.

When he looked back at the previous year, the pattern was obvious:

- $1,900 property taxes (due twice a year)
- $1,200 auto insurance (annual)
- $1,100 car repairs and maintenance (lumpy)
- $800 travel for family obligations (unpredictable timing)
- $600 medical/dental out-of-pocket (random)

None of those were freak accidents. They were life.

He had been absorbing them with a credit card "just for a month," then paying it down, then getting hit again. The balance never exploded, but it never fully disappeared either. It became a low-grade fever in his finances. He was constantly paying interest for the privilege of being surprised by predictable things.

He created four sinking funds:

- *Taxes*: $160/month
- *Insurance*: $100/month
- *Car*: $100/month
- *Family Travel*: $75/month

Total: $435/month.

That number felt aggressive. So he made one more decision that turned out to be the key: he didn't try to "find" $435 by being good. He created it by redirecting money that was already leaving his life:

- $120 from canceling and downgrading subscriptions he didn't truly use
- $90 from negotiating his internet and phone plan
- $75 from a realistic reduction in restaurants (not zero, just fewer)
- $150 from a temporary pause on extra principal payments to a low-interest student loan (he was still paying on time; he just shifted the *extra* until the cash-flow spikes were smoothed)

Within six months, the change was visible. Not because his life got cheaper, but because his money got calmer. When the semi-annual tax bill came, he paid it without using credit. When his dad needed a last-minute flight, the travel fund covered most of it and the remainder came from his guilt-free spending instead of his retirement contributions.

His finances didn't become perfect. They became less fragile. That's the point.

Where to keep sinking funds so you don't ``accidentally" spend them

A sinking fund fails for one of two reasons:

1. You didn't fund it enough.
2. You funded it, but it sat in the same account as your everyday spending and got eaten by convenience.

Make it slightly harder to touch than your spending money, but not so hard that you'll avoid using it when it's appropriate.

Practical options:

- *Separate savings account*(s) at your bank, ideally labeled by purpose.
- *One "sinking funds" savings account* with a simple tracking note (a small table in your notes app that says: Car $X, Home $Y, Gifts $Z). This is low-tech and very effective.
- *Multiple buckets within one account* if your bank offers sub-accounts or "vaults."

The best setup is the one you will actually maintain. If opening multiple accounts feels like a hassle, don't. Use one account and a clean tracking note. The purpose is to separate the money mentally and operationally from daily spending.

One more behavioral detail: give the account a name that makes you hesitate before stealing from it. "Sinking Funds" is fine. "Not a Credit Card Payment" is better.

What happens when the fund isn't ready yet

Early on, your sinking funds will be underbuilt. That's normal. The first year is the awkward year.

When an irregular expense hits before the fund is full, use a simple rule:

- Pay what you can from the sinking fund.
- Cover the remainder from guilt-free spending if possible (tighten up for a few weeks).
- If you must use a credit card, treat it like a bridge, not a new lifestyle: create a payoff plan immediately, and keep contributing to the sinking fund so the same bill doesn't hit you the same way next time.

This is how you break the cycle. Not with shame, but with a plan that gets a little sturdier each month.

Also: when the bill is paid, the fund does *not* go back to zero emotionally. You don't say "we failed." You say "we used the system." That mental reframing matters. The purpose of a sinking fund is to be spent.

How sinking funds protect your emergency fund (and your peace)

A real emergency fund is sacred because it buys you time when life truly goes off-script. Without sinking funds, your emergency fund becomes a junk drawer. You keep opening it for batteries and tape and random screws until, one day, you need a flashlight and it's empty.

Sinking funds keep the emergency fund clean. They handle the predictable weirdness so the emergency fund can handle the unpredictable disaster.

And there's a quiet emotional benefit: your financial life stops feeling like an endless ambush. When you know where the money is coming from, you can focus on the decision itself—Which mechanic? Which flight? Which dentist?—instead of the background panic of "How will we pay for this?"

That reduction in stress is not a luxury. It's a performance upgrade. Calm people make better decisions.

A simple, concrete takeaway you can do in 30 minutes

Open your banking app and do three things:

1. Create one new savings account named *Irregular Bills*.
2. Pick your first three sinking funds: *Car*, *Home*, and *Gifts & Holidays* cover a shocking amount of real life.
3. Set one automatic transfer for your next payday—even $25 per fund. The amount can be small. The habit cannot.

Then, the next time a "surprise" bill shows up, pay at least part of it from that account on purpose. That single act teaches your brain a new story: the bills aren't emergencies anymore, and you're not the kind of person who has to borrow money from the future just to survive the calendar.

2.3 Automating Your Catch-Up Plan

The most expensive financial mistake isn't a bad stock pick or one splurge weekend. It's assuming your future will be funded by your *best intentions*. Intentions don't survive real life. Real life is a Thursday at 9:47 p.m. when you're tired, your inbox is loud, your kid needs something for tomorrow, and you've already made 200 tiny decisions since breakfast. That's not the moment to negotiate with yourself about retirement contributions.

Automation is how you get your plan done on your worst days, not just your best ones.

People sometimes resist this idea because it sounds cold—like handing your money to a robot and hoping for the best. But the opposite is true. A well-designed automated system is deeply personal. It reflects what you value, protects you from predictable chaos, and quietly builds momentum while you're busy being a person.

There's also a reason automation works that has nothing to do with personality. Researchers have found that when retirement savings are set up as the default—automatic enrollment, automatic payroll deductions—participation rises dramatically, and many people stick with the default settings for a long time. In plain English: if savings happens *without you having to re-decide it*, it tends to keep happening. If it requires monthly willpower, it tends to fade.

The real goal: make the ``good" choice the *easy* choice

A time-crunched household doesn't need more information. It needs fewer decision points.

Think about how money normally behaves in checking: it arrives, sits there like a buffet, and gets eaten by whatever shows up first—bills, errands, convenience spending, the "oh right" expenses you forgot. By the end of the month, you look back and wonder why there's never anything left for catching up.

Automation changes the order of operations. Instead of hoping you'll save what's left, you decide what you're saving *first*, and then the rest of your life adjusts around what remains. It's reverse budgeting with a motor attached.

The goal isn't to automate everything. The goal is to automate the few actions that create the biggest long-term difference:

- retirement contributions
- emergency and sinking funds (so irregular expenses don't become credit card balances)
- on-time bill payment (so your credit and cash flow stop taking unnecessary hits)

If those happen by default, the rest of your budget becomes dramatically easier to live with.

The Hub-and-Spoke system: one account to rule the chaos

Here's the simplest architecture that works for most people: a *Hub* account receives income and routes it out automatically to a few *Spokes*. You don't need a dozen accounts. You need a clean flow.

The hub is usually a checking account—the one your paycheck lands in. The spokes are purpose-built destinations:

- *Investments* (401(k)/403(b)/457 payroll contributions, IRA contributions, brokerage if appropriate)
- *Fixed Bills* (mortgage/rent, utilities, insurance, minimum debt payments)
- *Spending Card* (a debit card or credit card used for guilt-free spending and variable expenses)

The power move is that money leaves the hub on a schedule you control. You don't wait until you "feel ready." You design the riverbed so the water flows the same way every time it rains.

A strong Hub-and-Spoke system does three things:

1. It protects Future You from being outvoted by Present You.
2. It makes bills boring (boring is a financial compliment).
3. It gives you a clear spending boundary without tracking every purchase.

Make the system safe: guardrails before speed

Automation should never feel like a trap. The fear people have—"What if I overdraft?"—is valid. The solution is not to avoid automation; it's to build in guardrails.

Start with these basics:

- Keep a small buffer in the hub (even $300–$1,000) so timing glitches don't cause overdrafts.
- Align transfer dates with paydays and bill due dates. Don't guess. Look at the calendar once, then set it.
- Use autopay for the bills that matter most (housing, utilities, insurance, minimum debt payments), but stay cautious with bills that vary wildly unless you have a buffer.

And remember: you're allowed to stop or change preauthorized transfers. Banks generally provide mechanisms to cancel or stop future debits; you're not signing away control. Automation is a tool, not a vow.

Where catch-up really happens: payroll beats willpower

If you have access to a workplace retirement plan, payroll deduction is the cleanest automation available. It's invisible, consistent, and it never waits for motivation.

For readers approaching 50, catch-up contributions are a major lever: in many employer plans, people who are age 50 or older by year-end can contribute extra beyond the standard elective deferral limit, subject to IRS rules and the plan's features. The details matter (plan types differ, and catch-up generally applies after you've hit the regular limit), but the practical takeaway is simple: once you're eligible, your retirement plan may let you save more through payroll without changing your lifestyle month to month.

There's a second, newer wrinkle: SECURE 2.0 introduced rules that can require certain higher-wage earners to make catch-up contributions as Roth (after-tax) rather than pre-tax, and the

IRS has issued guidance and timelines that affect when employers must operationalize those changes. That's not a reason to delay saving; it's a reason to check your paystub and plan elections carefully and confirm what your employer's payroll system is actually doing.

If your employer offers a "Save More Tomorrow"-style feature (automatic escalation of contributions with raises), take it seriously. It's one of the few financial hacks that uses human nature instead of fighting it. If you increase your contribution when your salary rises, you're less likely to feel the pinch—because you never got used to spending the entire raise.

The automation map (Hub-and-Spoke flowchart)

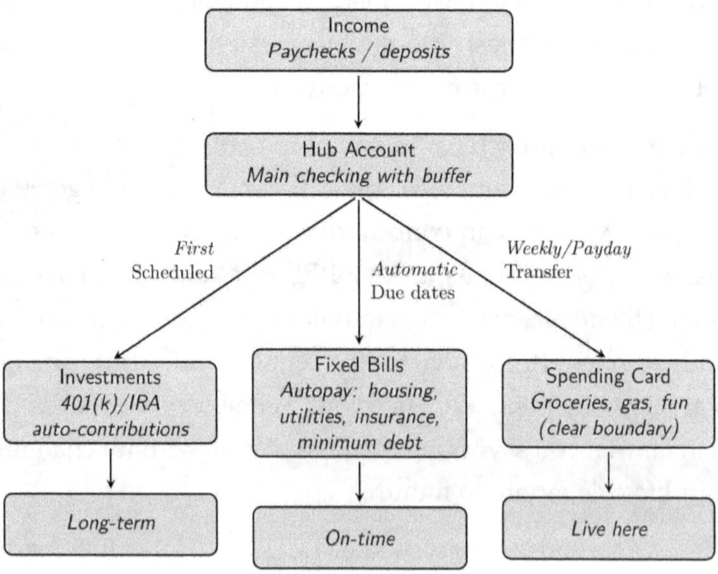

A concrete setup that works even if you hate ``systems"

Here's a workflow that's simple enough to run without becoming your new hobby:

1. *Set retirement contributions through payroll first.* If your employer matches, aim to capture the full match as a baseline. Then decide on your catch-up ramp: either a fixed increase today or an automatic increase each time you get a raise.
2. *Automate sinking funds next.* A single transfer on payday into a savings account labeled "Irregular Bills" can prevent dozens of credit card moments later.
3. *Put fixed bills on autopay from the hub.* Choose either the due date (if your cash flow is stable) or a consistent day after payday.
4. *Move your spending boundary out of your main checking account.* Use a dedicated spending card (credit or debit) and pay it from the hub on a schedule. This turns "How are we doing?" into a glance at one balance.

This works because it respects two truths:

- You will not out-discipline a messy cash-flow design.
- You don't need to micromanage spending when you've already protected the priorities.

Case study: Serena, 49, high income, low margin

Serena makes $145,000 a year and still felt broke. That sentence annoys people until they live it. Her issue wasn't shopping sprees; it was timing and leakage. She had a mortgage, student loans, two kids in activities, and a habit of paying for irregular expenses with a credit card "for points," then carrying the balance longer than she intended.

She wanted to catch up hard before turning 50, but every month ended the same way: she would *intend* to contribute to her IRA, then the money would get eaten by a mix of bills and life.

We rebuilt her flow, not her personality.

Serena's automated plan looked like this:

- 401(k): increased payroll contribution by 2% immediately, then set an auto-escalation of 1% at each annual raise.
- Hub buffer: kept $1,000 minimum in the hub account at all times.
- Payday+1: automatic transfer of $350 to "Irregular Bills" savings.
- Payday+2: autopay of the spending card set to a fixed amount ($1,400) rather than "statement balance," to avoid draining the hub unpredictably.
- Quarterly: scheduled calendar reminders to review contribution limits and confirm paycheck withholding was doing what she thought it was doing.

The results were not glamorous, but they were powerful. Her credit card stopped being a bridge to the future and went back to being a convenience tool. She stopped "forgetting" to invest because the money no longer waited for her attention. Most importantly, she said something that matters more than a spreadsheet: "I feel like my money finally has adults in charge."

Common automation mistakes (and how to avoid them)

Mistake: Automating too much, too fast. If you set up ten transfers and three new accounts in one weekend, you'll spend the next month untangling timing issues. Start with one or two automations, let them run for two pay cycles, then add the next.

Mistake: Using autopay without cash-flow visibility. Autopay is wonderful until it triggers an overdraft. Avoid the problem by keeping the hub buffer, aligning dates, and setting alerts for low balances.

Mistake: "Whatever is left" investing. Leftover investing is how most people stay behind. Choose a number, automate it, and force your lifestyle to negotiate with the remainder.

Mistake: Forgetting taxes on side income. If you have freelance or gig income, automation should include a tax spoke. Even a simple separate savings account where you skim a percentage of each payment can prevent an ugly surprise.

A small but potent tactic: automate the decision after a raise

Raises are rare moments when you can change your savings rate with minimal pain. The money isn't yet committed to your lifestyle, so redirecting part of it is easier than cutting later.

Try this rule: when you get a raise, increase your retirement contribution by half the raise (or at least 1%). Let the rest show up in your paycheck so it still feels rewarding. This isn't about austerity; it's about using timing to your advantage.

If you're eligible for catch-up contributions (generally at age 50 by year-end for many plans), that raise moment becomes even more valuable. Your payroll system can do the heavy lifting, but you have to tell it what to do.

The ending that matters: a plan that runs while you sleep

A catch-up plan succeeds when it stops depending on your mood. The win isn't "I followed the budget." The win is opening your accounts six months from now and noticing, almost with surprise, that Future You has been getting paid every single paycheck.

Pick one automation to set up this week: either a payroll increase to your retirement plan, or a payday transfer into an "Irregular Bills" sinking fund. Do it while you're calm, so you don't have to be calm later.

2.4 Debt Triage and Credit Health

Debt is one of those words that can make smart, capable adults feel instantly smaller. Not because they don't understand math, but because debt has a way of turning math into mood. You can be handling a career, a family, and a calendar packed with obligations—and still feel a jolt of shame when you open a credit card app.

That shame is useless. Debt isn't a personality test. It's a contract with a price tag. Once you see how the price tag works, you can decide what to keep, what to kill, and what to refinance into something less predatory.

And credit health? That's not a vanity metric. It's a discount you earn on the cost of borrowing. A stronger credit profile can mean lower interest on a mortgage, a car loan, even some insurance premiums in many states. It can also mean the difference between getting approved and getting shut out. You don't need to obsess over a perfect score. You need to protect the handful of factors that actually move the needle.

Debt triage: stop treating every balance like it has equal urgency

Triage is what emergency rooms do when too many people arrive at once. They don't treat everyone in the order they walked in. They treat the bleeding first.

Household debt works the same way. If you have multiple balances—credit cards, a personal loan, a car loan, student loans, maybe a medical payment plan—your job isn't to feel bad about all of it. Your job is to identify the debt that is actively siphoning your future and plug that leak first.

A simple way to sort debt is by how it behaves:

- *Wealth-destroying debt*: high interest, revolving, and easy to re-borrow (credit cards carrying balances, payday loans, many personal loans). This is the debt that charges you rent for being human.
- *Neutral debt*: moderate interest with a clear payoff schedule and an asset or credential on the other side (some student loans, some auto loans). Not fun, but not always a five-alarm fire.

- *Strategic debt*: low interest and manageable within your cash flow, often tied to long-term assets (a reasonably priced mortgage). This isn't *good* debt in a moral sense; it's debt that may not deserve your first extra dollar compared to retirement contributions or wiping out 24% APR card balances.

This is how you stop spinning your wheels. The point is not to hate debt. The point is to stop paying premium prices for it.

The mechanics that make debt feel like quicksand

Interest is simple in theory: you borrow money and pay a fee to use it. The part most people don't internalize is how the *timing* of interest turns into a long, expensive tail.

Credit cards are the most common offender because they're revolving: there's no finish line unless you create one. If you carry a balance, interest typically accrues daily and gets added to what you owe. The bigger the balance, the more interest you're charged. The more interest you're charged, the harder it is to reduce the balance. That loop is why credit card debt feels insulting: you're paying money to stay in the same place.

Minimum payments make the loop worse. The minimum is designed to keep you current, not to get you free. Early in repayment, a large portion of your minimum can go to interest, which means your balance barely moves. The lender gets paid reliably; you get a long horizon of "still dealing with this."

The emotional experience of this is brutal: you're doing what the statement asks, but the problem doesn't shrink fast enough to reward you. That's why debt payoff needs strategy, not just effort.

A quick diagnostic: where is your interest bleeding you?

If you want one high-impact habit, it's this: once a month, look at your statements and find the line that says *interest charged.*

Not the balance. Not the minimum payment. The interest.

That number is the cost of delay. It's also the easiest way to rank which debt deserves your next extra dollar. A $6,000 balance at 27% APR is not the same as a $6,000 balance at 5%. The APR is the gravity.

If you don't have the statements handy, use a rough rule: any debt above 10% APR deserves scrutiny; any debt above 15% is an emergency in slow motion. You can still live your life while you pay it off, but you should treat it like a leak you can't ignore.

Credit score demystified: what matters and what's mostly noise

Credit scores are often discussed like mystical numbers that judge you. They're not mystical. They're predictive models: probability estimates of whether you'll repay as agreed.

The model has a few big levers. The commonly cited breakdown for FICO scores goes roughly like this: payment history is the largest factor (often cited around 35%), and amounts owed (including credit utilization) is next (often cited around 30%). The other factors—length of history, new credit, and credit mix—matter, but they tend to be secondary for most people.

In practice, that means you can stop chasing hacks and focus on what actually saves you money:

- *Pay on time.* This is the heavyweight. A single 30-day late mark can do real damage, and it can linger on your reports for years.
- *Keep revolving balances low relative to limits.* This is utilization. If your cards are maxed out, your score can drop even if you've never missed a payment. The model reads high utilization as "stress."

Everything else is seasoning. Useful, but not the main dish.

One important nuance: you can be debt-free and still have credit. You don't need to carry a balance to build a score. Using a credit card and paying it off is typically enough to generate payment history. Interest is optional. On-time behavior is the point.

Credit utilization: the stealth factor most people accidentally sabotage

Utilization is one of the most frustrating parts of credit scoring because it can punish you for a totally normal month. Put a large expense on a card, your balance spikes, and if the statement closes before you pay it down, the reported utilization can jump.

A few ways to manage this without turning your life into a credit-score hobby:

- Pay the card down *before* the statement closing date if you know you have a temporary spike.
- Spread spending across more than one card (if you can manage it responsibly) so one card doesn't look maxed out.
- If your income and habits support it, request a credit limit increase. This can lower utilization without changing spending—but only do this if it won't tempt you into bigger balances.

This matters because better credit can lower borrowing costs precisely when you're trying to create margin. Lower interest on a refinance, a car loan, or a personal loan can free up monthly cash flow that can go to retirement contributions or debt payoff.

The unsexy superpower: checking your credit reports

People often confuse credit reports with credit scores. Your *report* is the raw file: accounts, limits, balances, payment history, and public records. Your *score* is a number calculated from that file.

If your report has errors, your score can suffer for no good reason. Errors happen. Sometimes it's a simple reporting mistake. Sometimes it's identity theft. Either way, the fix starts with looking.

In the U.S., the federally authorized place to get free copies of your credit reports is AnnualCreditReport.com, as referenced by the Federal Trade Commission's consumer guidance. When you pull your reports, you're not hunting for a perfect number. You're scanning for landmines:

- accounts you don't recognize
- late payments you believe are wrong
- incorrect balances or limits
- old negative items that should have aged off
- personal information that looks off (addresses you've never lived at)

Treat this like checking smoke detectors. You don't do it because you expect a fire. You do it because the consequences of not checking are expensive.

A case study: Nora's "fine" debt that quietly blocked her options

Nora is 43 and earns $92,000. She didn't consider herself "in debt," because her balances weren't enormous: about $9,500 across two credit cards, plus a car loan. She made payments on time. She wasn't getting collection calls. She was functioning.

But she was also stuck. Every month she paid $350—$450 toward the cards, and the balances barely moved. She kept telling herself she would "get aggressive" once work calmed down. Work didn't calm down.

Then she went to finance a used car after her old one died. The loan terms she was offered were worse than she expected, and the payment strained her budget. That was the moment she finally looked at her credit utilization. One card had a $10,000 limit with a $7,800 balance. The other had a $5,000 limit with a $1,700 balance. Even with on-time payments, her revolving utilization was high enough to drag her score down.

Her fix wasn't dramatic; it was targeted:

- She automated her minimum payments (so "busy month" could never create a late payment).
- She redirected one sinking fund temporarily: $200/month shifted from *Travel* into an extra card payment until utilization dropped.
- She made one mid-cycle payment each month, before the statement closed, to keep reported utilization lower.
- She called the issuer on her older card and requested a limit increase once she'd stabilized her balances.

Within a few months, her score improved enough to refinance the car loan at a meaningfully lower rate. The monthly savings went straight into continuing the credit card payoff. Nora didn't become a new person. She just stopped bleeding interest and stopped accidentally presenting as "financially stressed" to the scoring model.

That's the heart of credit health: it expands your options when you need them.

When debt collectors are involved: shift into documentation mode

If you're dealing with collections, the emotional temperature rises fast. The goal is to cool it down with process.

Consumer protections exist, and the Consumer Financial Protection Bureau (CFPB) provides guidance and sample letters for dealing with debt collectors. Two practical principles matter immediately:

- *Get everything in writing and keep records.* Dates, names, what was said, what you sent.
- *Ask for validation information and don't assume a caller is correct.* Mistakes and even scams happen.

You also have the right to request that a collector stop contacting you. That doesn't make the debt vanish, but it can stop harassment and give you breathing room to decide your next move rationally.

If you're in this situation, the priority isn't optimizing a credit score. The priority is stopping the chaos: confirm what's legitimate, negotiate if appropriate, and build a plan that prevents new delinquencies. One new late payment can be more damaging than you expect, and it's often preventable with automation and a small cash buffer.

Debt triage rules that keep you from making things worse

When people feel pressure, they sometimes do the financially equivalent of running on a sprained ankle: they can still move, but they're worsening the injury.

A few rules keep your progress clean:

- *Never miss a payment to accelerate another payoff.* Payment history is too valuable. Automate minimums on everything.

- *Don't drain your emergency buffer to zero.* A completely empty buffer forces you back to credit cards the next time the calendar throws a punch.
- *Stop adding new debt while paying off old debt.* This sounds obvious, but it often requires a practical change: use the sinking funds, adjust the spending boundary, or temporarily pause optional categories so you don't refill what you're trying to empty.

This is also where the Hub-and-Spoke setup shines. If your bills are being paid automatically and your irregular expenses have a home, your credit cards stop being the default shock absorber. That alone can change the trajectory.

A concrete takeaway: a 20-minute "credit and interest audit"

Block 20 minutes this week and do the following:

1. Pull your credit card statements and write down the APR and last month's interest charged for each card.
2. Circle the card with the highest APR *or* the highest monthly interest cost (often the same). That's the bleeding source.
3. Set autopay for minimum payments on every debt if you haven't already.
4. Get your credit reports from AnnualCreditReport.com and scan for errors or unfamiliar accounts.

Then make one move that reduces risk immediately: either an extra payment to the worst APR balance, or a mid-cycle payment to lower utilization before the statement closes. The point isn't to win in one month. The point is to stop paying unnecessary tolls for borrowing money—and to earn your way into cheaper borrowing when you truly need it.

2.5 Strategic Debt Elimination

Debt payoff advice tends to come in two flavors: moral lectures ("stop buying lattes") or math lectures ("optimize your interest rate spread"). Neither helps much when you're 44, you're carrying a few balances, and your life is already a full-contact sport.

The move that changes everything is smaller and more practical: pick a payoff strategy you can execute *reliably* for a long time. Debt is rarely defeated by one heroic month. It's defeated by a system that keeps working when you're tired, busy, irritated, or discouraged.

Two strategies dominate for a reason. They're both simple enough to run without a finance degree, and both can work brilliantly:

- *Debt Snowball*: pay off the smallest balance first for quick wins, then roll that payment into the next balance.
- *Debt Avalanche*: pay off the highest interest rate first to minimize total interest, then roll that payment into the next highest rate.

The best method is the one you won't quit.

Before you choose: the one rule that makes every method work

No matter what strategy you pick, one rule stays sacred:

> Pay the minimum on every debt, every month, on time.

That rule protects your credit, prevents late fees, and keeps a temporary rough month from turning into a long-term mess. It also buys you the right to focus your *extra* money—the margin you're creating—on one target at a time.

If you haven't automated minimum payments yet, do it. A missed payment can do more damage than an inefficient payoff order ever will.

Debt Snowball: momentum you can feel in your chest

The snowball method is designed for the part of you that's exhausted. It acknowledges something most personal finance articles pretend isn't true: motivation is a resource, and it runs out.

With the snowball, you list debts from smallest balance to largest. You throw every extra dollar at the smallest debt while paying minimums on the rest. When the smallest is gone, you take the entire payment you were making on it (minimum plus extra) and "roll" it onto the next smallest.

The magic is emotional: a balance disappears. That disappearance creates relief, then confidence, then a sense of speed. It's like finally clearing clutter off a floor—you can walk again.

Snowball is especially powerful when:

- you have several small-to-medium balances
- you're prone to getting discouraged
- debt feels psychologically heavy even when the numbers are manageable

You're not doing it because you can't understand interest rates. You're doing it because you understand *yourself*.

Debt Avalanche: the cold, beautiful efficiency of math

The avalanche method is designed for the part of you that hates waste.

With the avalanche, you list debts from highest interest rate to lowest. You throw every extra dollar at the highest APR debt first while paying minimums on the rest. When the top APR debt is paid, you roll the payment to the next-highest APR.

The benefit is financial: you generally pay less interest and often get out of debt faster *for the same monthly payment*. It's an efficiency engine.

Avalanche tends to be best when:

- your highest-interest balance is also one of your larger balances (so the math win is meaningful)
- you're motivated by optimization
- you can tolerate a longer stretch before the first "win" (a closed account)

If snowball is a psychological tailwind, avalanche is a shaved-off interest bill.

The side-by-side: order of operations you can picture

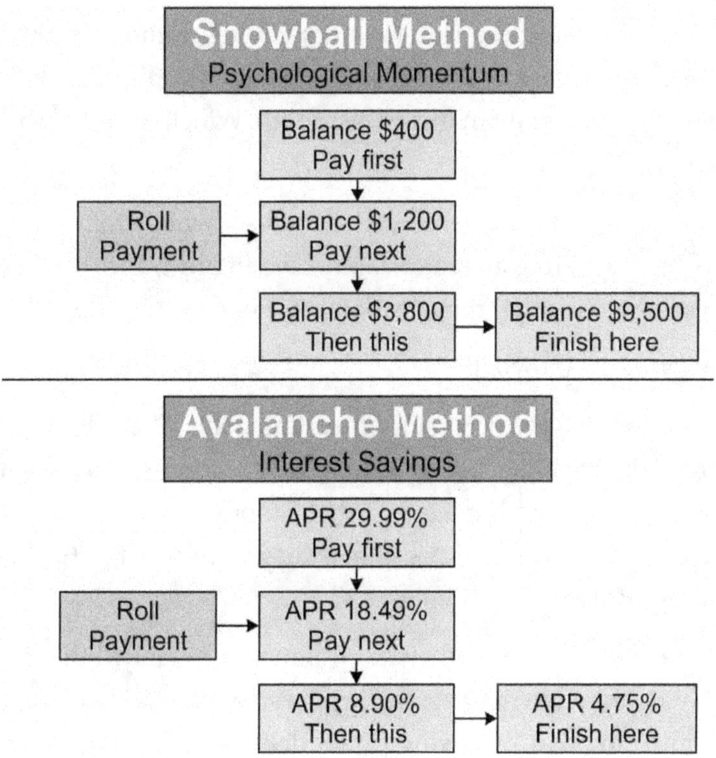

A real-world case: Tom and Elena choose the ``wrong'' method and win anyway

Tom (46) and Elena (45) had what many couples have: decent income, busy lives, and a debt pile that looked manageable until it didn't. Their balances:

- Card A: $1,100 at 27% APR
- Card B: $4,600 at 22% APR
- Personal loan: $7,800 at 11% APR
- Store card: $650 at 29% APR

They were doing the classic thing: paying minimums everywhere and occasionally throwing an extra $200 at whichever balance annoyed them most that month. Which meant nothing ever died.

Tom wanted avalanche. Elena wanted snowball. Tom argued (correctly) that avalanche would save interest. Elena argued (also correctly) that they needed a fast win or they'd burn out.

They picked snowball.

Within weeks, the $650 store card was gone. Then the $1,100 card. Two accounts closed in under two months. The psychological effect was immediate: they stopped feeling like they were drowning in four directions. They were now fighting one meaningful target at a time.

Was avalanche more efficient on paper? Probably. But here's what mattered in their actual life: the snowball method turned them from "sort of paying down debt" people into "we are paying off debt" people. That identity shift is not fluffy. It's the difference between an approach that lasts and an approach that gets abandoned in the first stressful season.

They didn't ignore math forever. Once they had only two debts left, they switched to avalanche—because at that point the choice felt energizing, not overwhelming. Strategy can evolve as your emotional bandwidth returns.

How to choose: one simple test

If you're stuck deciding, use this test:

> Which outcome will make you feel more powerful in 60 days: closing an account, or knowing you attacked the highest APR?

If the thought of closing an account makes you feel lighter and more in control, pick snowball.

If the thought of crushing the highest APR makes you feel satisfied—like you're refusing to be exploited—pick avalanche.

Both methods rely on the same backbone: consistent extra payments and not re-borrowing the freed-up money. The method just determines which target gets the spotlight.

The trap doors: promotional APRs and deferred interest

Debt payoff gets messy when promotional offers enter the room wearing a friendly face.

Balance transfers and promotional APR cards can be useful tools, but they come with sharp edges. The CFPB has warned that some promotional and deferred-interest offers can produce surprise interest charges if you don't meet the terms perfectly. Deferred interest, in particular, can be nasty: if you don't pay the balance in full by the deadline, you may owe interest retroactively.

A practical way to handle promos without getting burned:

- Put the promo end date in your calendar with two reminders: 60 days before and 30 days before.
- Know whether it's *0% APR* (interest waived during the promo) or *deferred interest* (interest accumulating in the background).
- If you're using a 0% transfer, treat the required payoff as a fixed monthly bill: balance ÷ months left.
- Avoid charging new purchases to a card carrying a promotional balance unless you understand how the grace period works. Some setups can cause new purchases to accrue interest immediately if you're not paying the statement balance in full.

Promos are not a payoff strategy. They're a tool that can support one. The strategy is still snowball or avalanche, plus consistent execution.

Automation with control: keep an off-ramp

Autopay and automatic transfers are powerful, but debt payoff can involve trial and error with timing. The point is to create a system you can adjust without chaos.

A few guardrails:

- Keep extra payments separate from minimum payments. Minimums should be automated and untouchable; extra payments can be paused in a true emergency without triggering late fees.

- Watch for payments that could cause overdrafts. A small buffer in your hub account is cheaper than repeated bank fees.
- Know your rights around preauthorized transfers. Under the Electronic Fund Transfer Act / Regulation E framework, consumers generally can stop payment of preauthorized electronic transfers with enough notice and can revoke authorization. That matters if a lender or service provider is pulling amounts you didn't expect or if you need to prevent repeated debits while you resolve a dispute.

Control reduces fear. Reduced fear makes automation sustainable. Sustainable automation is how debt payoff becomes inevitable instead of aspirational.

The ``roll-up" effect: why small wins turn into big speed

Both methods rely on the same engine: each time you eliminate a debt, you *roll* its payment into the next target. That creates a compounding payment effect.

If you were paying $45 minimum on a small card and you were adding $155 extra, that's $200/month going to that debt. When it's gone, you don't "free up" $200 to spend. You redirect it to the next debt. The next debt gets paid faster, which frees up more payment, which accelerates the next payoff.

This is why debt payoff can feel slow and then suddenly fast. The first payoff is the hardest. Later, the same amount of income produces more progress because you're stacking payments onto fewer targets.

A concrete takeaway: set up your ``attack payment" today

Open your notes app and write:

- All debts + balances + APRs + minimum payments
- Your *attack payment*: the extra amount you can reliably pay each month (even $50 counts)
- The target debt (smallest balance for snowball; highest APR for avalanche)

Then do two actions:

1. Set minimum payments to autopay for every debt.
2. Schedule your attack payment to go out automatically 1–2 days after payday.

When the first balance hits zero, take a screenshot. Not for social media. For your own brain. Proof changes behavior. And behavior—more than any trick—is what turns debt from a permanent background noise into a problem with an end date.

2.6 The Income Accelerator

Cutting expenses can feel like trying to lose weight by exhaling harder. Yes, you can tighten a few things. Yes, leaks matter. But there's a hard ceiling on how far "spend less" can take you when you're already paying for housing, food, transportation, and the basic dignity of not living on panic.

Income is different. Income is an open field.

That's why the most underrated midlife financial strategy isn't a clever budget category or a new app. It's offense. It's using your experience—the stuff you've earned the hard way—to get paid more per hour, more per project, or more per year. When retirement feels too close, offense buys you time and options faster than penny-pinching ever will.

This isn't about turning your life into hustle culture. It's about acknowledging a practical truth: many households end up needing some form of extra income at some point, and a meaningful number of adults do gig or side work in real life. The question isn't whether side income exists. It's whether you can build it in a way that doesn't wreck your health, your taxes, or your core career.

The midlife advantage nobody advertises

At 25, you might have had energy and ambition. At 45, you have something rarer: *pattern recognition*. You know how work actually gets done. You know where projects stall, where clients get anxious, where managers need reassurance, where quality matters, where speed matters, and where perfection is a waste of time.

That knowledge is monetizable.

The mistake is thinking income growth is only for people who love networking or who are naturally assertive. It's not. Income growth is a set of moves, many of which are quiet and strategic:

- negotiating for more in your existing role
- repositioning your skills into a higher-paying lane
- adding a side stream that fits your life instead of consuming it

The goal is not to do all three. The goal is to find the one that gives you the best return on your time.

Start with a simple question: is your problem rate, hours, or volatility?

Before you chase money, diagnose what kind of income problem you have.

Rate problem: you're underpaid for what you do. *Hours problem*: you're capped on hours because of caregiving, health, or a demanding job, so you need more money per hour. *Volatility problem*: income is inconsistent (commission, self-employment, seasonal work), and the unpredictability is the stress.

Each problem points to a different solution:

- Rate problems respond to negotiation and job changes.
- Hours problems respond to higher-value work, consulting, or skills that command premium pricing.
- Volatility problems respond to stabilizing cash flow (retainers, predictable shifts, or building a buffer and tax system).

This keeps you from doing the common midlife thing: adding a side hustle that pays too little, taxes you too much (emotionally), and collapses in three months.

Negotiation: the highest-paying hour you can work

If you can tolerate a little discomfort, salary negotiation is often the most profitable "side hustle" available. It pays you for every hour you work going forward.

The problem is that many people approach negotiation like a confession: "I work hard and I've been here a while." That's not a business case. Longevity is nice, but it doesn't automatically translate to revenue, risk reduction, or operational leverage—the things employers actually pay for.

A stronger approach is to build a *one-page case* that answers three questions:

1. *What problems do you reliably solve?* (Not tasks. Problems.)
2. *What outcomes improved because you were there?* (Time saved, errors reduced, revenue supported, clients retained, projects delivered.)
3. *What would break if you left?* (This is not a threat; it's a reality check about dependency and replacement cost.)

Then you ask for a specific number, not a vibe. If naming a number makes you sweat, that's normal. The point is to give the conversation a target.

A practical script that doesn't feel theatrical:

> "I want to talk about aligning my compensation with the scope I'm covering. Over the last *X* months, I've delivered *A, B,* and *C*. I'm now owning *D* that used to sit at a higher level. I'd like to move my base salary to $*N*."

If the answer is no, you haven't failed. You've gathered information. Then the follow-up question becomes the real lever:

> "What would need to be true for that number to be a yes?"

You're looking for a path, a timeline, or a clear "not here." All three are useful.

A case study: Angela's $18,000 raise without a new degree

Angela is 41 and works in a healthcare-adjacent operations role. She thought of herself as "not technical," which in her mind meant she didn't have access to higher pay. Meanwhile she was quietly doing the work that kept the place from falling apart: scheduling, vendor coordination, fixing process glitches, training new hires, calming down upset patients, and catching errors before they became compliance problems.

Her income had been creeping up slowly, and she was trying to close a retirement gap by being aggressively frugal. It wasn't working; her fixed costs were real.

Instead of hunting for a second job, she did a two-week audit of her own value:

- documented recurring problems she prevented (missed appointments, billing errors, staffing gaps)
- quantified time saved ("this used to take 3 hours, now it takes 45 minutes")
- collected two short emails from internal stakeholders praising her work (social proof without being weird about it)

She asked for a salary adjustment. Her manager couldn't do the full number immediately but offered a title change plus a smaller raise, with a review in six months tied to specific metrics.

Angela took it—and used the title change as leverage to interview elsewhere. Within three months she landed a new role at another organization with an $18,000 increase. Same basic skill set. Different lane. Her retirement contributions went up automatically through payroll. She didn't become a different person; she just stopped underselling a competence she'd been treating as "just being responsible."

That's the midlife advantage: you already have skills. The work is packaging and pricing.

Career pivots: smaller than you think, faster than you fear

The word "pivot" can sound like quitting your job to become a different species. Most income-boosting pivots are more like moving one seat over.

A good pivot has three traits:

- it uses your existing experience as an unfair advantage
- it pays meaningfully more or creates clearer growth
- it doesn't require you to torch your life to make it happen

One of the clearest midlife patterns in the labor market is that some fields grow and some compress. Broad projections often show continued demand in healthcare and certain skilled roles, while some routine administrative work gets pressured by automation and software. You don't need to become an economist to respond intelligently. You just need to look honestly at your current lane and ask: *is my role becoming more valuable, or more replaceable?*

If it's becoming more replaceable, the answer isn't panic. It's proactive positioning:

- move closer to revenue (sales support, client success, implementation)
- move closer to risk (compliance, quality, security, regulated processes)
- move closer to complexity (project management, systems, cross-functional coordination)

Complexity is where experienced adults get paid. Not because complexity is fun, but because it's hard to automate the part where humans coordinate, persuade, and make judgment calls.

Side income: choose low-overhead, high-control options

Side income is common and often useful, but it comes with two midlife constraints: time and energy. The best side income for someone over 40 is rarely the most glamorous. It's the one that doesn't explode your schedule or create a second set of obligations that you resent.

A helpful way to choose is to score ideas on three dimensions:

- *Overhead*: money required upfront (equipment, inventory, advertising)
- *Control*: can you say no, set your hours, pause when life gets heavy?
- *Rate*: what do you earn per hour after expenses and taxes?

Many popular side hustles look good before you do the math. Driving, delivery, and task platforms can be legitimate bridges, especially in a tight season, but they can also deliver low net pay once you account for vehicle wear, downtime, and taxes. Reselling can work if you already have an eye for value and a system, but it can turn into "stuff management" if you don't.

The most powerful side income for many midlife earners is *consulting or contracting* in their current skill set:

- bookkeeping for small businesses
- project coordination
- coaching and tutoring in a subject you already know
- writing/editing in an industry niche
- IT support or training for non-technical teams
- HR and recruiting support
- notary work, if demand exists locally

These can be started small, with one client, and expanded only if the return on time is worth it.

The tax reality of side income: don't let a good hustle become a bad surprise

Side income has a predictable trap: you get paid, you feel richer, and then tax season shows up and demands its cut.

In the U.S., gig income is generally taxable. If you're self-employed, you may owe not only income tax but also self-employment tax, and there are thresholds and rules that can surprise people who are new to it. A common bright line is that net earnings from self-employment of $400 or more generally triggers self-employment tax filing obligations. The details vary by situation, but the practical move is consistent: *set money aside as you earn it*.

A simple system:

- Open a separate savings account called *Taxes*.
- Each time you get paid, transfer a percentage immediately (many people start with 25–30% as a rough placeholder, then adjust once they see their actual tax situation).
- Keep basic records: income received, expenses, mileage if relevant, and receipts.

This does two things: it prevents a tax emergency and it helps you evaluate whether the side hustle is actually profitable. Money you haven't reserved for taxes isn't fully yours.

Fraud and scams: the income accelerator's dark side

When money feels tight, people become more vulnerable to offers that promise speed. Scammers understand that urgency makes brains skip steps.

If an income opportunity requires you to pay upfront, move money through unusual channels, accept overpayments, buy gift cards, "verify" accounts with codes, or share sensitive information quickly, assume it's a trap until proven otherwise. Plenty of legitimate side income exists; none of it requires you to bypass common sense.

A simple rule that saves pain: if the offer pressures you to act before you can verify details, it's not an opportunity. It's a test.

Where the extra money should go first (so it actually changes your life)

Extra income is only powerful if it lands somewhere purposeful. Otherwise it gets absorbed into lifestyle and vanishes like every other raise you've ever had.

A clean order of operations for the first dollars of extra income:

- build or strengthen your emergency fund (so you stop borrowing from the future)
- fund sinking funds (so irregular expenses stop creating credit card balances)
- eliminate high-interest debt (so you stop paying the "time tax")
- increase retirement contributions (so Future You compounds)

You don't have to do all of these at once. The point is that income should create margin, not just nicer months.

A concrete takeaway: design a 30-day income experiment

You don't need a grand plan. You need an experiment with a finish line.

For the next 30 days, pick one of these tracks:

1. *Negotiation track*: write your one-page case, schedule the conversation, and ask for a number.

2. *Pivot track*: apply to five roles that are one seat over but pay more, using your current experience as the headline.
3. *Side-income track*: choose one low-overhead service you can offer, set a simple price, and try to land one client.

Then add the rule that makes the experiment matter: any extra money earned goes automatically into one designated destination—a high-interest debt payoff, an "Irregular Bills" sinking fund, or a retirement contribution. When income gets a job, it stops disappearing. And when it stops disappearing, the retirement gap starts shrinking in a way you can actually feel in your daily life.

Chapter 3

Securing the Foundation: Safety Nets and Housing

Before we hit the accelerator on your investments, we need to check the brakes. When you are in your 20s, a financial disaster is a learning experience; when you are in your 40s or 50s, it can be a permanent setback that forces you to work years longer than planned. This chapter isn't just about 'playing it safe'-it is about building a floor beneath your feet so you can take the necessary risks to grow your wealth above it. We will tackle the unsexy but critical triad of midlife financial security: liquidity, insurance, and the roof over your head.

3.1 Building a Bulletproof Emergency Fund

Three months of expenses sounds crisp. Responsible. Grown-up. It also sounds like the kind of advice that assumes your life is still shaped like your twenties: quick job searches, fewer people depending on you, and a body that bounces back after you "power through" an illness.

Midlife rarely cooperates with that script.

In your 40s and 50s, the financial shocks aren't just a flat tire and a surprise dental bill. They're the longer, uglier disruptions: a layoff that drags on because your role is specialized; a parent who needs help now, not later; a teenager who breaks an arm the week the car needs tires; a health issue that turns into a months-long stamina problem. The bills don't arrive one at a time. They pile up, then they start calling.

A bulletproof emergency fund isn't about hitting a generic number. It's about buying back *options*. Options to be picky about your next job instead of desperate. Options to keep your retirement contributions steady when life gets loud. Options to sleep like a person whose finances are not one calendar mishap away from a credit-card spiral.

Why ``three months" can fail when you're older (and more valuable)

The standard advice usually smuggles in three hidden assumptions:

1. *Your income can be replaced quickly.*
2. *Your expenses can be cut quickly.*
3. *Your shock will be a one-time bill, not a long stretch.*

All three are shakier in midlife.

First, replacing income. The higher your compensation and the narrower your expertise, the longer the hiring loop tends to be. It's not just getting an offer; it's getting the *right* offer—one that matches your level and doesn't permanently downshift your earning power. A mid-career job search can involve multiple interview rounds, background checks, start dates that land a month out, and a slow ramp before bonuses or commissions return. You can be doing everything right and still be stuck in a waiting room for longer than three months.

Second, cutting expenses. When you're 26, "cut spending" can mean swapping restaurants for groceries and pausing a few subscriptions. When you're 46, the expense structure is heavier and more contractual: mortgages, car payments, insurance premiums, kids' activities, therapy, orthodontics, student loans (yours or theirs), and support for family members. Many of those bills don't politely shrink just because your employer did.

Third, the shape of emergencies. The costliest emergencies are often *duration problems*. A medical issue that keeps you from traveling for work. A caregiving situation that forces you to reduce hours. A company reorg that turns into six months of uncertainty, then a severance package with strings attached. The emergency fund that works in midlife is designed for time, not just surprise invoices.

And here's the quietest reason three months fails: midlife is when you finally have something meaningful to protect. Retirement accounts with real momentum. Home equity you'd rather not liquidate in a panic. A reputation you don't want to damage by taking the first job that will have you. Three months can be enough to prevent an overdraft. It may not be enough to prevent a bad decision.

The real goal: your ``Sleep Well" number

Forget "months." Start with a more human metric: *How much cash would let me handle a serious disruption without making a mistake I'll regret for ten years?*

That amount is your *Sleep Well* number. It's not a moral scorecard. It's a personal engineering problem with inputs that are specific to your life.

To calculate it, you'll build two things:

- A *baseline runway*: the minimum needed to keep the lights on and avoid financial damage.
- A *volatility buffer*: extra runway based on how likely you are to face a long interruption and how hard it would be to shrink your expenses.

This is where generic advice becomes expensive. If you work in a stable role with high demand, a two-income household, and flexible spending, your Sleep Well number might be smaller than you think. If you're a single earner in a cyclical industry with a big mortgage and a specialized title, it may need to be meaningfully larger.

A good emergency fund is not "as big as possible." It's *big enough to change your behavior in a crisis*. If your fund is too small, you won't act calmly. If it's too large, you may hoard cash that should be invested for long-term goals. The point is the *right* size.

Step 1: calculate your baseline runway (the bills that don't care)

Start with your essential monthly outflow. Not your ideal budget. Not your "we'll just eat beans" fantasy. The real number that keeps your life intact.

A clean way to do this is to list your non-negotiables in three buckets:

- *Shelter*: rent or mortgage, property tax/HOA if applicable, basic utilities.
- *Keep-the-machine-running*: groceries, transportation, insurance premiums, minimum debt payments, basic phone/internet.
- *Human needs*: medications, therapy, childcare you can't instantly replace, essential costs tied to work (licenses, required travel not reimbursed).

Two rules make this honest:

1. Use actual numbers from the last three months of spending (bank/credit card statements), not guesses.
2. Include insurance premiums even if your employer currently pays them. If you lose your job, health insurance doesn't become optional; it becomes *urgent*.

Your baseline runway is typically *three to six months of essential outflow*. Why the range? Because essential outflow is only half the story. The other half is how likely you are to face an income gap that lasts longer than you want.

Step 2: add a volatility buffer (because job loss isn't random for everyone)

Now you weight the risk. Not in a doom-and-gloom way—in the same way you'd price insurance or decide whether to buy the extended warranty on a laptop (which, by the way, you probably shouldn't).

Ask yourself four questions and score them with brutal honesty:

1. *Industry volatility:* Are layoffs common in your field? Is hiring seasonal? Do companies treat your function as a "nice to have" when budgets tighten?
2. *Role specificity:* Can you move laterally into many companies, or is your job title rare and your interviews unusually specialized?
3. *Household flexibility:* Is there a second income that can cover essentials? Could you pick up short-term work quickly without derailing caregiving needs?
4. *Expense rigidity:* How much of your monthly outflow is contractual (mortgage, car, tuition, support obligations) versus adjustable (restaurants, travel, hobbies)?

If you score "high" on two or more, your volatility buffer should be meaningful: often another *three to six months* of essential outflow.

This is the part many midlife professionals resist because it feels like admitting vulnerability. But here's the twist: the more valuable your income is, the more expensive it is to lose it for a year. A larger emergency fund is not pessimism. It's leverage.

A concrete example: the six-month layoff that turned into nine

Consider Maya, 48, a product marketing leader at a mid-sized tech company. Household of four. One spouse worked part-time, mostly for flexibility with school pickups and an aging parent.

Maya's initial emergency fund goal was three months of expenses, which in her mind meant "enough to cover the mortgage." She had about $18,000 saved. It felt respectable. It also assumed the emergency would be a single event and her next paycheck would arrive on schedule.

Then her company announced a reorg. Her role was eliminated. She received eight weeks of severance—a cushion that looked generous until she realized how hiring works at her level.

The job search timeline went like this:

- Month 1: sorting out health insurance, updating resume, networking, emotional whiplash, first recruiter calls.
- Month 2: interviews begin, but many roles pause for "budget review."
- Months 3—4: several multi-round processes; one role disappears after the final interview.
- Month 5: a strong lead, but the offer is delayed by internal approvals.
- Month 6: offer arrives—start date four weeks out.
- Months 7—8: income returns, but bonus/commission structure won't pay until later.

Maya did many things right. She didn't stop. She networked. She stayed disciplined. The timeline still stretched.

Her essential outflow was $7,400/month, once she included COBRA health insurance, groceries, utilities, car costs, minimum debt payments, and the expenses tied to supporting her parent. Three months of essentials was $22,200—already above her savings.

The difference got financed the usual way: credit cards "temporarily," then a 401(k) loan, then pausing retirement contributions and telling herself she'd catch up later. That story is common, not because people are irresponsible, but because their emergency fund was built for a smaller life.

What would bulletproof have looked like for Maya?

- Baseline runway: 6 months of essentials = $44,400.
- Volatility buffer: tech role + specialized leadership hiring + expense rigidity = add 3 more months = $22,200.

Total Sleep Well number: about $66,600.

That number can sound outrageous until you compare it to the cost of the alternative: retirement contributions paused for a year, credit card interest, and the permanent loss of compounding on money that never got invested. The emergency wasn't the layoff; it was the financial decisions she was forced to make under time pressure.

The most overlooked part: your deductible is part of your emergency fund

People often treat insurance and emergency funds as separate topics. In real life, they're welded together.

Your emergency fund must be large enough to handle the *deductibles* you've chosen: home, auto, health. A $5,000 home insurance deductible is not a theoretical number; it's the price of admission to get help when your roof leaks or a pipe bursts. A high-deductible health plan can create a multi-thousand-dollar cash need at exactly the wrong time.

A bulletproof fund accounts for this with a simple add-on:

$$\text{Deductible Reserve} = \max(\text{home deductible}, \text{auto deductible}) + \text{health out-of-pocket exposure you could face quickly}$$

You don't need to fully fund the worst-case annual maximum out of pocket for every scenario. But you do want enough liquidity that you don't turn a claim into debt.

If you want to raise deductibles to lower premiums (often a smart move), do it in the same week you increase cash reserves. Otherwise you're not reducing risk; you're transferring it.

Where to keep it: the three-tier system (so the money is usable)

An emergency fund has one job: be there when the emergency shows up. That means two things:

- It must be safe (not subject to market swings).

- It must be liquid (accessible without penalties or delays).

This is why retirement accounts are a poor primary emergency fund. They can be tapped, but that's different from being designed for tapping. Taxes, penalties, paperwork, and opportunity cost turn "available" into "expensive." A bulletproof emergency fund keeps you from raiding the future to pay for the present.

The practical approach is to separate your cash into three layers:

1. *Checking* (*1 month*): your operating cash. This prevents overdrafts and late fees. It's not an emergency fund; it's your household's bloodstream.
2. *Buffer* (*1 month*): a slightly larger cushion in a savings account you can move instantly. This is for the annoying surprises: car repair, travel for a family issue, an insurance deductible.
3. *Deep reserves* (*3—12 months*): the real emergency vault. This is for job loss, major medical disruption, or a long income gap.

If this feels fussy, remember what you're really doing: you're preventing one problem from turning into five problems. When money is layered correctly, the small shocks don't force you to touch the deep reserves. And when you do need the deep reserves, you can access them without selling investments at a bad time.

For most readers, a high-yield savings account or money market deposit account at an insured bank or credit union works well for the buffer and deep reserves. The yield matters, but accessibility matters more. A slightly lower interest rate is a small price to pay for immediate control when life goes sideways.

A midlife twist: include a ``job search budget" line item

If you're building your emergency fund around job loss risk, include job-search costs in your planning. These aren't extravagant; they're realistic:

- Higher health insurance premiums during transitions.
- Travel or networking expenses.
- Training or certification costs that make you employable faster.
- A modest "sanity budget" so you don't cut every joy and then burn out.

The sanity budget is not indulgence. It's fuel. When you're under stress, the temptation is to clamp down so hard that you rebound into impulsive spending later. A small planned release valve keeps you steady.

A practical way to do this is to calculate your essential outflow, then add 5%—10% as a job-search stability margin. That margin is often what keeps you from making the classic mistake: taking a bad-fit job just to stop the bleeding.

How to build it fast without feeling like you're suffocating

Building a large emergency fund in midlife can feel like trying to fill a swimming pool with a kitchen faucet—especially if you're also catching up on retirement. The goal isn't to sacrifice everything; it's to build momentum.

Three tactics work because they're behavioral as much as mathematical:

1) Automate the minimum, then sweep the extra.

Set a baseline automatic transfer on payday (even $100). Then add a monthly "sweep" rule: any cash above a certain checking balance moves to the emergency fund. This captures irregular inflows (bonuses, reimbursements, uneven spending months) without requiring willpower.

2) Use "found money" with a plan.

Tax refunds, bonuses, gifted cash, and even the savings from paying off a car can disappear into lifestyle inflation if you don't assign them a job. A simple rule is:

$$\text{Found Money Split} = 50\%\ \text{emergency fund},\ 50\%\ \text{highest-priority goal}$$

If your emergency fund is dangerously low, tip the split temporarily. The point is that you're directing windfalls with intention, not guilt.

3) Shrink the fund-building pain by defining "enough."

A bulletproof emergency fund is not an endless project. Define your Sleep Well number, and define an interim milestone—often one month of essentials in checking + one month in savings. That early win reduces anxiety quickly, which makes it easier to keep going.

A warning that sounds boring until it saves you: keep it separate

The emergency fund should be psychologically protected. If it's sitting in the same account you use for vacations, it will quietly become the vacation fund. If it's in the same brokerage account as your investments, it will feel like "available cash" and get deployed into the market at the first dip.

Give the emergency vault a different bank, a different nickname, and a different mental label. "Emergency" is fine, but something like "Job-Loss Runway" or "Family Shield" is better because it's harder to rationalize away.

One practical trick: keep the deep reserves in an account without a debit card. Not because you want friction for emergencies, but because you want friction for *non-emergencies*. A one-day transfer delay is a nuisance when you want new furniture and a blessing when you're tempted to treat the fund like a flexible spending account.

Actionable takeaway: calculate your number this week, then fund it in layers

Write down three numbers and you'll have a plan that fits your real life:

1. *Essential monthly outflow* (from the last three months of statements).

2. *Target runway* (6 months if your income is stable; 9—12 months if your industry/role/household setup makes long gaps plausible).
3. *Deductible reserve* (the cash you'd need to confidently file an insurance claim without going into debt).

Then implement the layered structure immediately: one month in checking, one month in a fast-access savings buffer, and the rest in deep reserves. Once the first two layers are in place, you'll feel the difference in your body before you see it on a spreadsheet—because the goal was never to win a personal-finance purity contest. The goal was to make it harder for a bad month to become a bad decade.

3.2 Insurance: Protecting Your Earning Power

If someone asked you, on the spot, to list your biggest assets, you might name the house, the cars, maybe the 401(k). That's normal. Those are the things you can point to, insure, and screenshot.

But the asset that quietly funds every other asset is the one most people under-protect: your ability to earn a paycheck.

In midlife, this isn't abstract. You're close enough to retirement to feel the countdown, but far enough away that you still have a long stretch of working years left—often the highest-paid years you'll ever have. A long interruption doesn't just make one year hard; it can knock the whole plan off its rails. When contributions stop, compounding doesn't politely pause. It keeps going without you.

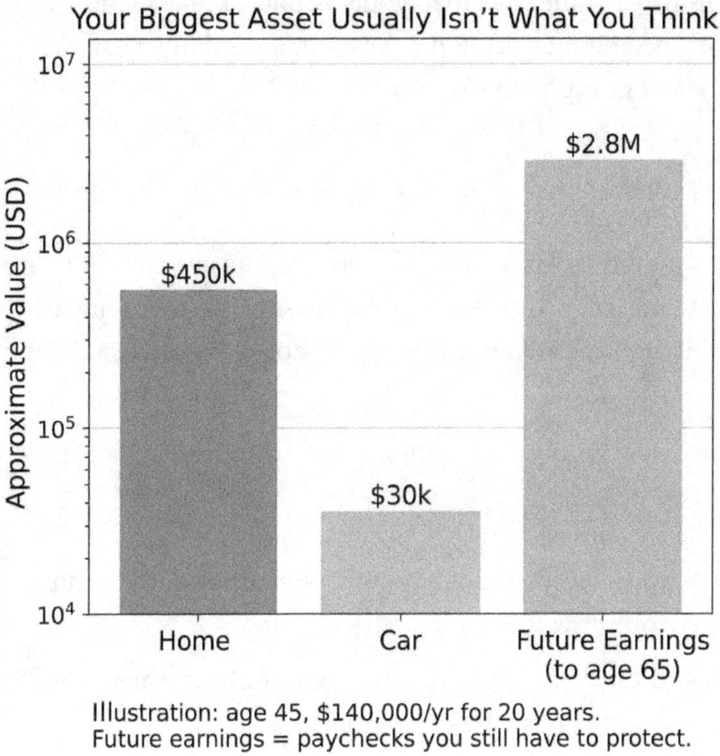

That chart is intentionally a little rude.

Even if you adjust every number—smaller salary, later retirement, a cheaper home—the shape of the truth stays the same: future earnings are a towering asset. It's the engine that pays the mortgage, funds the retirement accounts, covers health insurance, keeps the household stable, and gives you choices. If you're going to buy protection, the engine deserves first dibs.

Insurance is how you transfer a risk you can't afford to carry. The mistake most households make is transferring the risks that feel vivid (a cracked windshield, a stolen laptop) while self-insuring the risks that would actually break them (a multi-year income interruption, a lawsuit, a disability that changes what work is possible).

The mindset shift: stop insuring objects; insure outcomes

"Insure the car" is really shorthand for "insure my ability to replace the car." "Insure the home" is shorthand for "insure my ability to keep living my life if the house is damaged." Those are outcomes.

Once you think in outcomes, your priorities rearrange themselves:

- If an event would require you to raid retirement accounts, you want it insured.
- If an event would force a home sale on someone else's schedule, you want it insured.
- If an event would make you dependent on family or debt, you want it insured.

This isn't fear-based. It's a grown-up version of what you already do when you build an emergency fund: you keep small problems small. Insurance is how you keep the *big* problems from becoming your identity for a decade.

The most ignored gap: disability insurance (because it's unglamorous and inconvenient)

Life insurance is discussed at dinner parties. Home insurance gets attention after storms. Disability insurance barely gets mentioned, even though it protects the asset that dwarfs the rest.

Disability insurance replaces income if you can't work due to illness or injury. Not just "can't lift heavy things," but "can't reliably perform the material duties of your occupation." That definition matters. A surgeon with a hand tremor and a sales leader with severe migraines are both dealing with real disabilities, even if they can still *technically* do some kind of work.

Long-term disability (often abbreviated LTD) is the policy that keeps a health event from turning into a retirement disaster. If you only take one action after reading this, make it this: *find your LTD coverage and read the details like you're reading a contract. Because you are.*

Why the gap hits hardest after 40:

- You're more likely to have a health event that is not instantly fixable.
- Your compensation is higher, so a partial replacement hurts more in dollars.
- Your retirement timeline is shorter, so you have fewer years to recover.
- Your financial obligations tend to be larger and less flexible.

A disabling event doesn't have to be dramatic. It can be the slow kind: back issues that limit sitting, autoimmune flares that crush energy, cancer treatment that changes stamina for a year, a mental health crisis that makes high-stress work temporarily impossible. Those are precisely the situations where "just grind through it" stops working.

A real-world case study: the injury that didn't look like a financial emergency (until it did)

Take Eric, 52, a project manager in construction with a solid salary and a strong work ethic. He'd insured everything he could touch: trucks, tools, home, even a small boat. He had a tidy emergency fund and felt proud of it.

Then he fell from a ladder and suffered a spinal injury. It wasn't a headline catastrophe. He wasn't paralyzed. He just couldn't stand for long periods, couldn't drive long distances, and couldn't tolerate the physical demands of job sites. He assumed he'd be back "soon."

Soon turned into months.

His employer had LTD coverage, which sounded comforting until the details mattered:

- The policy replaced 60% of salary, but only up to a monthly cap.
- Benefits didn't begin until after an elimination period (a waiting period) that lasted 90 days.
- Because the employer paid the premium, the benefits were taxable, shrinking the take-home amount.

Eric discovered that "60%" can mean "far less than you need." The mortgage didn't shrink 40%. Health insurance premiums went up. Physical therapy copays appeared like clockwork. His emergency fund bridged the waiting period, but after that he was living on a much smaller, taxed benefit while trying to heal.

The financial damage wasn't bankruptcy. It was quieter: retirement contributions stopped, credit cards floated expenses, and stress slowed recovery. A year later he was working again, but the lost year of savings and the debt hangover stretched into the future.

Two insights come from Eric's story:

1. Emergency funds and disability insurance are a relay team. Cash gets you through the waiting period; LTD keeps the long disruption from draining everything.
2. Coverage details are not fine print. They *are* the product.

How to audit your workplace LTD in 20 minutes (without becoming an insurance nerd)

Most midlife professionals have some LTD through work and assume that means they're covered. You may be. Or you may have a nice-sounding policy that collapses at the exact moment you need it.

Pull your benefits summary and look for these items:

Replacement rate and the monthly cap.

A typical plan might say "60% of base salary." Then it adds a cap like "up to $10,000 per month." If you're a higher earner, the cap is the real replacement rate. Do the math with your actual pay.

What counts as earnings.

Is it base salary only, or does it include bonus, commission, overtime? Many plans cover base pay only. If a meaningful chunk of your income is variable, you can be underinsured without realizing it.

Elimination period (waiting period).

Common elimination periods are 90 or 180 days. This number should directly influence your emergency fund runway. If your elimination period is 180 days and you've built only three months of cash, you've created a gap on paper.

Tax treatment.

If you pay the premium with after-tax dollars, benefits are generally tax-free. If your employer pays, benefits are generally taxable. The same "60%" benefit can feel wildly different depending on taxes.

Definition of disability.

This is a big one:

- *Own-occupation* coverage generally means you're disabled if you can't do *your* occupation.
- *Any-occupation* coverage becomes stricter, often after a period of time, meaning you may be considered not disabled if you can do *any* reasonable job.

The difference between those definitions is the difference between protecting your career and protecting you from absolute destitution. The first preserves earning power; the second is closer to a safety net.

Benefit duration.

Some plans pay to age 65; some pay for a fixed number of years. If the benefit stops after five years and you're disabled at 52, you need to understand what happens next.

If you discover gaps, you have options: supplemental LTD through work (often offered during open enrollment), or an individual disability policy. Individual policies can be more expensive, but they can also be more tailored and portable if you change jobs. The goal isn't to buy the fanciest policy; it's to avoid the specific failure modes that would force you to raid retirement accounts.

The liability risks that can wreck your paycheck even if you're healthy

Not all threats to earning power come from health. Some come from the legal system. This isn't a pleasant dinner topic, but it's worth being honest: a lawsuit can become a second job you didn't apply for, and it can put your assets on the table.

That's where an umbrella policy enters, wearing its most boring name.

An umbrella policy provides additional liability coverage beyond your home and auto policies. Think of it as "catastrophe coverage" for the kind of event that produces a large claim: a serious auto accident where someone is badly injured, a guest injured on your property, a teenager's driving mistake, a dog bite, a defamation claim, a liability incident tied to rental property. You don't need to predict the exact scenario; you just need to recognize that the dollar amounts can get big faster than most people intuit.

Umbrella insurance is often relatively inexpensive compared to the amount of coverage, but it usually requires you to carry certain minimum liability limits on your auto and home policies. That requirement is not a nuisance; it's a clue. It's the insurance industry telling you where the pressure points are.

If you have meaningful assets, a higher income, a teenager who drives, a pool, a dog, frequent entertaining, or you simply want to be less exposed to one terrible day, an umbrella policy is "cheap sleep." It doesn't make you invincible. It makes you less likely to spend your 50s paying for an accident with your 60s.

Homeowners insurance: the hidden link between housing and liquidity

Home insurance looks straightforward until you file a claim. Then you learn what you actually bought.

A standard homeowners policy typically bundles several types of coverage: the structure (the dwelling), your personal property, liability, and loss of use (help paying for temporary living expenses if the home is uninhabitable). The bundle is useful, but it can also create blind spots because you assume "home insurance" is one thing.

Two midlife realities make this especially important:

- Your home is likely your largest single balance-sheet item.
- The cash demands after a home event are immediate (deductibles, temporary housing gaps, repairs not covered).

The emergency fund you built is what allows you to use insurance effectively. Without liquidity, you can end up under-repairing the home, taking on expensive debt, or delaying necessary work. It's a brutal irony: the people who need insurance most are often the people least able to float the out-of-pocket costs that come with using it.

Three practical things to check:

Replacement cost vs. actual cash value.

Replacement cost coverage aims to pay what it costs to replace or repair with similar materials today. Actual cash value generally subtracts depreciation. Two homes can have the same "coverage amount" on paper and wildly different claim experiences depending on this detail. If you want the policy to restore your life, not just offer a consolation prize, this is worth attention.

Your deductible is a cash decision.

A higher deductible lowers premiums, but it raises the amount of cash you must produce on short notice. If you choose a $5,000 deductible and you don't have $5,000 you can access immediately, you didn't save money; you accepted a risk you can't cash-flow.

What's excluded.

Many policies exclude floods and earthquakes and require separate coverage if you want protection there. The point isn't to panic-buy every add-on; it's to avoid being shocked by a denial when you assumed you were covered.

Auto insurance: boring until it protects your retirement plan

Auto insurance is the classic example of insuring a thing (the car) when what you really need to insure is the *liability* connected to the car. Collision and comprehensive matter, but in midlife, liability limits are often the bigger lever. If you have assets and income, low liability limits can be an invitation for your balance sheet to get dragged into someone else's medical bills.

A practical framing helps: if you caused a serious accident tomorrow, would your current auto liability limits cover the plausible costs? If the answer is "I'm not sure," that's not an answer; it's a to-do.

Umbrella coverage can sit on top of auto liability, but it doesn't replace the need for adequate underlying limits. Think of auto liability as the foundation and the umbrella as the extra floors.

The ``don't overinsure the small stuff" rule

Midlife catch-up plans are fragile in a specific way: you can't afford to waste money on protection that doesn't meaningfully reduce your biggest risks.

A useful rule of thumb: don't pay recurring premiums to insure a loss you could comfortably cover with your emergency fund. Save insurance capacity (and premium dollars) for the losses you *can't* absorb.

That doesn't mean choosing reckless deductibles or stripping coverage. It means being intentional:

- Raise deductibles only when your cash reserves can handle them.
- Prioritize liability and income protection over low-impact add-ons.
- Treat insurance as a system, not a pile of unrelated policies.

When you do this well, insurance stops feeling like an annoying expense and starts feeling like a strategic tool. You're not buying paper. You're buying the ability to keep making good decisions under stress.

Actionable takeaway: write your ``income protection scorecard" and fix one gap

Pick a quiet 30-minute window and write down these six facts in one place:

1. Your monthly essential expenses (the number your emergency fund was built around).
2. Your workplace LTD replacement rate *and* monthly cap.
3. Your LTD elimination period (90/180 days).
4. Whether LTD benefits would likely be taxable (who pays the premium).
5. Your auto liability limits and whether you have umbrella coverage.
6. Your homeowners deductible and whether you could pay it tomorrow without debt.

Then choose one fix you can execute in the next two weeks: increase liability limits, add an umbrella policy, enroll in supplemental LTD, or adjust deductibles to match your cash reality. The goal is not to create a perfect insurance portfolio. It's to make sure a single injury, accident, or claim can't hijack the next 20 years of paychecks you're still counting on.

3.3 Life Insurance for Midlife Families

The most emotionally honest way to think about life insurance is also the least comfortable: it's money you buy for people you love in a scenario you don't want to imagine.

In your 30s, the math is usually simple. You're protecting a young family against the worst timing imaginable: a mortgage that's still huge, kids who can't earn, and retirement accounts that are still seedlings. The typical advice—"get term insurance, plenty of it"—often fits.

In your 40s and 50s, the story changes. Not because you stop loving your family, but because the *job* of the policy changes. The kids are closer to independence. The mortgage is smaller (sometimes). Your investments are finally becoming real. At the exact same time, new pressures show up: college bills, aging parents, a spouse who might depend on your health insurance, and a retirement date that's close enough to feel personal.

This is the awkward middle. It's when people overpay because they're afraid to under-protect, and sales pitches are engineered to exploit exactly that feeling.

The midlife reality: your need for life insurance usually goes down while the price goes up

Life insurance need is often highest when your financial responsibilities are highest and your assets are lowest. As you build net worth, the need usually declines. Meanwhile, as you age, the cost of buying new coverage usually rises. That creates a squeeze: you feel like you should "lock something in" before it gets more expensive, even if the amount you actually need is shrinking.

This is where the industry loves to meet you with a permanent policy brochure and a calm voice that says, "You can't outlive it."

True. You also can't outlive a toaster, but you don't buy a premium toaster as a retirement plan.

The core question isn't whether a policy lasts forever. The question is: *what financial problem are you solving, and for how long does that problem exist?*

For most midlife families who are trying to close a retirement gap, the cleanest answer is still usually term insurance—because term insurance is designed to cover a *temporary* risk: the years when your household would be financially harmed by losing your income.

A simple model that makes the decision less emotional: ``Decreasing Responsibility"

Think of your life insurance need as a slope, not a cliff.

When your kids are small, your household's financial obligations are high. You're responsible for almost everything: income, childcare, stability, the long runway to adulthood. Over time, that responsibility decreases. Kids grow up. Debts get paid down. Retirement accounts grow. At some point, if you've built enough assets, your family becomes effectively *self-insured*—meaning your spouse (and anyone else who depends on you) could continue their life without a life insurance check.

That crossover point matters because it tells you the only truly important thing about life insurance in midlife: it's a *bridge*, not a monument.

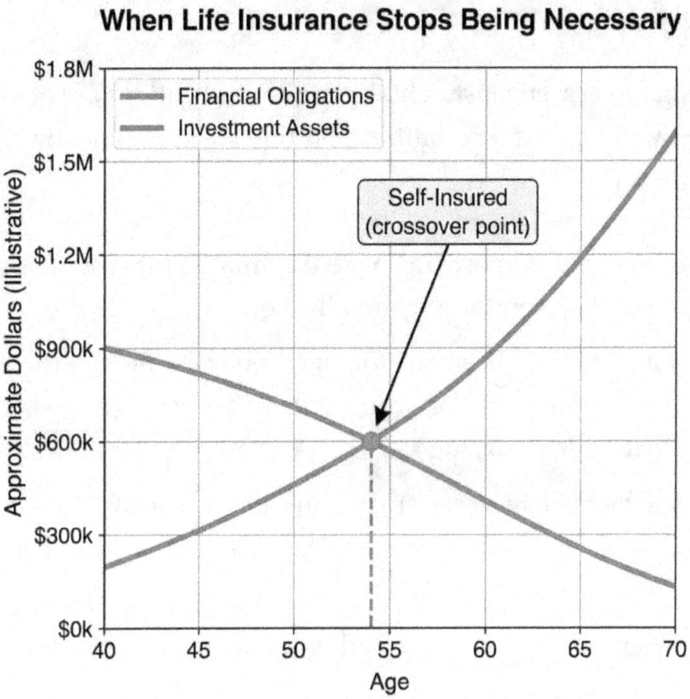

Curves are simplified to show the concept: obligations often fall as assets rise.

The precise shapes of those lines will differ for every family, but the intersection is the point. Before the crossover, life insurance is protecting people you love from a math problem they didn't create. After the crossover, the policy is often just an expensive habit.

What ``obligations" actually means (and why people underestimate it)

Obligations aren't just debt. In midlife, obligations are often a messy mix of responsibilities that don't show up neatly on a credit report:

- Remaining mortgage or rent commitments your spouse couldn't comfortably handle alone.
- Income replacement for day-to-day living (food, utilities, transportation) during the years your spouse is still working or raising kids.
- Child-related costs that don't disappear at 18: college support, health insurance gaps, or a young adult who needs a launchpad.
- Support for parents or other family members.
- One-time "settlement" costs: funeral expenses, legal fees, moving costs, paying off small debts that are emotionally draining to manage.

If you want a practical definition: obligations are the dollars your family would need to buy time and stability.

What ``assets" means (and why people overestimate it)

Assets, in this context, are resources that would actually be available to your survivors.

Your 401(k) balance is an asset, but it comes with rules and taxes. Home equity is an asset, but it may require a sale, a refinance, or a hard lifestyle change to access. A pension survivor benefit is an asset, but only if you elected it correctly. Social Security survivor benefits can be an asset, but they depend on eligibility and timing.

This is why you can't just glance at net worth and declare victory. The question isn't "Do we have money?" It's "Would this money be available soon enough, cleanly enough, and in a form that keeps the household stable?"

A case study: the 10-year term that saved a family from over-buying

Tanya and Luis were 47 and 49, with one child finishing high school and another in middle school. They had a 20-year term policy they bought when the kids were toddlers. It was nearing the end of its term, and the renewal quote was painful. A friend suggested whole life: "You'll build cash value and never have to worry again."

They almost did it. The pitch hit every midlife nerve: fear of outliving coverage, anxiety about retirement, desire to leave something behind.

Instead, they did a simple bridge calculation.

The problem they needed to solve: If either died in the next decade, the survivor would struggle to cover the mortgage and college costs *without* draining retirement accounts. After about ten years, the mortgage would be much smaller, the youngest child would be near independence, and retirement accounts would be materially larger.

So they bought a 10-year term policy sized to cover that specific gap. Not forever. Just long enough to get to their crossover point.

The result was not just lower premiums. It was emotional clarity. They stopped shopping for "the perfect policy" and started buying a tool that matched a deadline.

A decade-long bridge won because it was honest about what they were protecting: not eternity, just the fragile years when losing one income would force bad financial decisions.

The practical question: how much term insurance is enough now?

A clean midlife approach is to size coverage around the specific outcomes you want to guarantee. Here are examples of outcome-based targets:

- "If I die, my spouse can pay off the mortgage" (or pay it down to a manageable payment).
- "If I die, my spouse can fund the remaining years until the kids are independent without touching retirement accounts."
- "If I die, my spouse can cover college up to a specific amount, not an unlimited blank check."

Then subtract resources that would realistically be available: existing savings, employer life insurance, expected survivor income sources, and the portion of investments you would be comfortable having your spouse use.

This is where many people discover they don't need as much new insurance as they assume. They need *targeted* insurance.

Don't forget survivor income (but don't build your whole plan on it)

For many U.S. households, Social Security survivor benefits can be meaningful. In broad terms, a surviving spouse may become eligible for survivor benefits starting at age 60 (or earlier if disabled), and minor children may be eligible until adulthood (with some extensions in specific situations). Those benefits can reduce the amount of life insurance you need.

But there's a reason planners are cautious about leaning too heavily on them: eligibility rules, timing, and benefit amounts vary, and the household still has to survive the early period after a death when paperwork is happening and income can be disrupted.

A useful middle ground is to treat survivor benefits as a stabilizer, not the main pillar. Let life insurance handle the "right now" needs (cash flow, mortgage, childcare), and let survivor benefits be the supportive background income that reduces pressure over time.

The whole life sales pitch: what it gets right, and what it gets wrong

Whole life (and many other forms of permanent life insurance) is often sold as a two-for-one: protection plus investment.

The part it gets right: permanence can be valuable when the need truly is lifelong (for example, certain estate planning goals, a lifelong dependent, or specific business planning situations). Some people also value the forced savings aspect and the predictability.

The part it gets wrong for many midlife catch-up savers: it often solves the *wrong* problem at a high cost.

If your primary goal is to close a retirement gap, you need two things:

- adequate protection during your highest-responsibility years, and
- maximum dollars flowing into flexible, transparent, low-cost investment vehicles.

Permanent policies can be expensive, and the early years are typically front-loaded with costs. The cash value may grow, but it often takes time before it behaves the way the illustration implies. In other words, it can tie up money you would prefer to deploy aggressively toward retirement accounts, debt payoff, or simply maintaining flexibility.

A blunt but helpful sentence: *if you need life insurance, buy life insurance; if you need investments, buy investments.*

Keeping them separate makes it easier to compare costs, understand liquidity, and change course if your life changes.

Red flags that should slow you down before buying permanent insurance

If you hear any of the following, pause and ask for clarity in writing:

- "It's just like a Roth IRA." (It's not. Different rules, costs, and tax treatment.)
- "You can borrow from it anytime, tax-free." (Loans have mechanics and consequences; it's not the same as free money.)
- "This is the only way to build wealth safely." (No legitimate financial plan requires one product as the sole path.)
- "Term insurance is throwing money away." (Term is renting protection. If you needed protection, the rent did its job.)

None of this is a moral judgment on permanent insurance. It's a warning about context. Midlife families who are behind on retirement savings often need affordability and flexibility more than they need a complex policy with a long break-even.

The unsexy risk that ruins good plans: nobody can find the policy

Even the best life insurance policy fails if your family can't locate it when they need it.

This is more common than people think, especially when policies were bought decades ago, employers changed, mail was ignored, and paperwork lived in a drawer that got cleaned out during a move.

A practical midlife move is to create a one-page "If I Die" sheet:

- insurer name, policy number, and phone number

- beneficiary designations (and where they're documented)
- employer benefits contact info
- where the physical/ digital documents live

Then tell your spouse (and ideally one trusted person outside the household) where that sheet is. This is not being dramatic. This is being kind.

Actionable takeaway: buy a bridge, not a souvenir

Do three things this month:

1. *Draw your crossover lines*: estimate your remaining obligations over the next 10–20 years and your investable assets trajectory. You're looking for the age when your family becomes effectively self-insured.
2. *Audit what you already have*: term policies, employer-provided life insurance, and beneficiary designations. Many people are either unknowingly underinsured or paying for overlapping coverage.
3. *If there's a gap, fill it with term insurance sized to a deadline*: often a final 10-year (or 15-year) term that protects the fragile years until your crossover point.

When you match the policy term to the years your family truly needs protection, life insurance stops feeling like a bet against yourself. It becomes what it was always meant to be: a time-limited promise that the people you love won't have to solve a financial puzzle while they're grieving.

3.4 The Housing Dilemma: Rent, Buy, or Hold?

Housing is the financial decision that refuses to stay in its lane. It's where spreadsheets collide with marriages, where the best "deal" can still feel like a terrible life, and where one signature can lock in your monthly cash flow for thirty years. By midlife, the question usually isn't "Should I buy a house someday?" It's sharper and more immediate: *Do I keep this mortgage into retirement? Do I downsize? Do I rent on purpose? Or do I stretch for a new place and hope my future self forgives me?*

The pressure is amplified because you can see retirement now. Maybe not clearly, but it's no longer a fuzzy concept. You can count the working years left on your hands without running out of fingers. That changes how housing risk feels. A big housing mistake at 28 is recoverable; a big housing mistake at 48 can become the plot.

The hard truth is that there isn't a universally correct answer. There is, however, a universally correct *method*: stop debating housing as an identity and start treating it as a cash-flow machine with a lifestyle overlay. When you do that, the fog lifts. Not because the decision becomes easy, but because it becomes honest.

The invisible cost of ``I can afford the payment''

People rarely go broke from the mortgage payment alone. They go broke from the *total cost of ownership*—the money that leaks out of the house in dozens of small, non-negotiable ways.

A payment is a headline. Ownership is the full article.

Owning tends to come with recurring costs that are easy to ignore when times are good:

- repairs and maintenance (the roof does not care about your budget)
- property taxes (which can rise even when your salary doesn't)
- homeowners insurance (which can jump after regional disasters or claims)
- HOA dues and special assessments (a polite term for "surprise bill")
- utilities that scale with square footage

In your 40s and 50s, these costs matter more because your margin matters more. Every extra $500/month in housing carrying costs is $500/month that can't be used to build retirement contributions, reduce debt, or keep your emergency reserves full enough to make you resilient. That $500 also has a psychological cost: it narrows your job flexibility. It makes you less able to quit a toxic role, take a pay cut for sanity, or withstand a layoff without panic.

A useful midlife question is: *Is my house making my life bigger, or making my options smaller?*

Renting isn't ``throwing money away"---it's buying flexibility

"Rent is throwing money away" is one of those phrases that sounds wise until you notice it's also a slogan.

Rent is not an investment. It's a service: shelter with a built-in exit ramp. When you rent, you're paying someone else to take on price risk, repair risk, and often insurance complexity. You're also paying for the ability to move without transaction costs that can feel like a minor surgery: agent commissions, closing costs, repairs to sell, moving costs, months of stress.

Flexibility has a price, but it also has value. In midlife, flexibility can be the difference between riding out a disruption and being forced into a bad financial choice.

Renting can be a strategic move when:

- your job situation is uncertain or highly mobile
- you're considering relocation for lower cost of living or family needs
- you're rebuilding finances and need predictable housing costs
- home prices (or mortgage rates) would force you into a house-poor scenario

It's also worth saying out loud: owning a home outright isn't the only way to feel secure. Many people feel more secure with a strong emergency fund, growing investments, and the ability to move if their health, career, or family changes.

Buying isn't just buying a house; it's buying a set of future constraints

Homeownership is often described as "forced savings" because a portion of your payment can go toward principal—the part of the loan balance you pay down over time. That's real. It can be a powerful wealth builder.

But forced savings has a personality. It's illiquid (hard to access without borrowing or selling). It's concentrated (one big asset in one market). And it can be fragile if your cash flow is tight.

By midlife, the key question is whether buying (or upgrading) helps you build wealth without starving the rest of your plan. If the house purchase crowds out retirement contributions, leaves you without reserves, and forces you to carry high-interest debt, it's not forced savings. It's forced stress.

When buying makes sense:

- you expect to stay long enough to justify transaction costs
- the payment plus ownership costs fit comfortably alongside retirement saving
- you have reserves for repairs and deductibles without touching retirement accounts
- the home supports your life in a way you'd pay for anyway (school district, proximity to family, stability)

When buying becomes a risk:

- you're stretching for the payment under the assumption that income will rise
- you're draining cash reserves to make the down payment
- you're ignoring maintenance because "we'll handle it later"
- you're counting on appreciation to make the deal work

Appreciation is dessert, not dinner. If the house only works if prices rise, it doesn't work.

The ``hold" decision: the underrated superpower of a boring, affordable mortgage

A lot of midlife homeowners are sitting on something that has become oddly valuable: an old mortgage with a low interest rate. It may not feel like an asset the way a brokerage account does, but it can function like one because it reduces your fixed monthly obligations. Lower fixed obligations increase your ability to save, invest, and survive shocks.

This is why "just upgrade" can be such an expensive impulse. The upgrade often comes with:

- a larger principal balance
- a higher interest rate than your current loan
- higher property taxes and insurance (because the property value is higher)

In other words, you trade a stable platform for a higher monthly break-even point. That might still be the right decision for your family, but you should treat it as a major strategic move, not a lifestyle reward.

Holding can be a wealth-building move if it allows you to:

- increase retirement contributions materially (not by $50, but by enough to matter)
- maintain robust cash reserves so you don't fear job disruption

- avoid taking on new debt for renovations or furnishing

If holding your current home is the difference between retiring with breathing room and retiring with a knot in your stomach, "boring" becomes beautiful.

The retirement question everyone avoids: do you want a mortgage in retirement?

A mortgage in retirement is not automatically bad. It's a tool: you're paying for housing over time instead of all at once.

The problem isn't the mortgage; it's the cash-flow fragility.

In retirement (or semi-retirement), your income is often less flexible. You can't easily work more hours to cover a surprise property tax increase. You may be withdrawing from investments, which makes market downturns more dangerous. A fixed mortgage payment can be manageable if it's small relative to your income and you have reserves. It can be suffocating if it forces large withdrawals during bad markets.

Here's a practical way to frame it:

> If markets dropped 25% the year you retired, would you still be okay making your housing payments without panic-selling investments?

That question pulls housing into the same risk-management frame as everything else. It's not about being debt-free as a moral achievement. It's about building a retirement that can take a punch.

Run the numbers with two lenses: monthly cash flow and ``escape velocity"

Most people run a rent-versus-buy comparison by comparing the mortgage payment to rent. That's like comparing the sticker price of a car to the cost of owning it for a decade.

In midlife, you need two lenses:

Lens 1: Monthly cash flow. Add up the true monthly cost of owning:

$$\text{True Monthly Ownership Cost} = \text{P\&I} + \text{taxes} + \text{insurance} + \text{HOA} + \text{maintenance reserve}$$

P&I is principal and interest. The maintenance reserve is the line item people skip because it feels fake until the water heater dies. A common planning range is 1%–2% of home value per year, but older homes or high-cost labor markets can demand more. The exact percentage matters less than the discipline of budgeting for it.

Then compare that total to rent for a comparable lifestyle (not necessarily the same square footage, but the same life: commute, safety, schools, walkability).

Lens 2: Escape velocity. Ask: which option increases your ability to build wealth *and* reduces the chance that a disruption will derail you?

Escape velocity is the point where your savings and investments start growing faster than your obligations. It's when your financial life begins to feel less like pushing a boulder and more like steering a vehicle.

Housing choices can either help you reach escape velocity or keep you stuck, even with a good income.

A concrete case study: the ``upgrade" that delayed retirement (and the alternative that didn't)

Jordan and Sam were 44 and 46 with two kids. They owned a house with a manageable mortgage and a rate they'd locked in years ago. The house wasn't perfect: smaller kitchen, no dedicated office, a backyard that sloped at an annoying angle. They were tired of it. Their friends were upgrading. Social media was full of "dream home" renovations.

They considered buying a larger home in the same school district. The new mortgage payment would be about $1,200/month higher, and property taxes plus insurance would add another $500/month. Total increase: roughly $1,700/month.

Jordan said the sentence that changes a lot of lives: "We can afford it."

They could—in the way that many people can afford things in their 40s: by assuming nothing goes wrong.

Then they ran the escape velocity lens. They were behind on retirement contributions and wanted to close the gap. That extra $1,700/month, invested consistently, could have been the difference between "hope" and "plan." It also would have been their buffer against job disruption, a health event, or a parent needing support.

They chose a third option: hold the house and do a targeted renovation that created a workable home office and improved storage. The renovation wasn't free, but it was finite. The new mortgage wasn't finite; it would have been a new permanent baseline.

Two years later, they had:

- increased retirement contributions enough to feel real progress
- kept an emergency fund that could cover a job-search period
- a home that functioned better without becoming a larger financial obligation

Their friends' upgrades looked prettier. Jordan and Sam's balance sheet looked calmer. Calm is a retirement strategy.

When downsizing is actually a wealth move (and when it's a mirage)

Downsizing is often sold as a simple equation: sell big house, buy smaller house, pocket the difference, retire early.

Sometimes it works exactly like that. Sometimes it doesn't, because downsizing has costs people underestimate:

- transaction costs (agent commissions, closing costs)
- moving costs (which can be shockingly high)
- repairs and updates needed to sell
- the cost of buying into a desirable smaller-home market (where "small" is expensive)
- higher interest rate if you replace a low-rate mortgage with a new one

Downsizing becomes a wealth accelerator when it produces *investable* cash and reduces ongoing costs. If it merely swaps one expensive market for another, or if the smaller home comes with high HOA fees and rising insurance premiums, it may not accelerate anything.

A quick truth test:

> After all fees and moving costs, will the move create at least one of the following: a meaningfully lower monthly cost, a meaningfully larger investment balance, or a meaningfully lower stress level?

If the answer is no, the move may be a lifestyle choice (which is valid), but it's not a wealth strategy.

Insurance and hazard costs: the housing line items that can swing the math

Homeownership costs aren't stable. Taxes and insurance can rise, sometimes sharply, and the increases often arrive with very little drama until they hit your escrow payment. In some areas, insurance has become a serious part of the housing affordability problem.

Two practical implications:

1) Your housing plan needs a "premium shock" margin. If your budget is so tight that a 20% increase in homeowners insurance would break it, you're not just house poor. You're *fragile*. Fragility is the enemy of a catch-up plan.

2) Coverage gaps can turn an event into a financial crater. Many standard homeowners policies exclude floods and earthquakes, and separate coverage may be required depending on where you live and your lender's requirements. The point isn't to memorize every exclusion; it's to recognize that hazard risk is part of the true cost of ownership. A house in a risky area isn't just a house. It's a bundle of costs and uncertainties.

Renters often don't see these costs directly, which is part of what they're paying for: someone else is absorbing the volatility.

The mortgage payoff vs. invest debate (through a midlife lens)

The classic question goes like this: "Should I pay off my mortgage early or invest the money?"

At midlife, there's a better version: *Which choice increases my ability to retire on time without panic?*

Paying down a mortgage can provide a guaranteed return equal to the interest rate you're avoiding. It also reduces your fixed monthly obligations, which can be psychologically powerful and practically useful.

Investing can provide higher expected returns over long periods, but with volatility. That volatility matters more when retirement is closer, because you have less time to recover from a downturn.

So the decision is less about being clever and more about matching your plan to your nervous system and your timeline.

A practical approach many people use:

- Keep investing consistently for retirement (because you can't eat a paid-off house).

- Use extra cash to pay down the mortgage only after emergency reserves are strong and retirement contributions are on track.
- If retirement is within a decade and the mortgage payment would create stress during market downturns, consider prioritizing principal paydown to reduce fixed costs.

The aim is balance: keep the wealth-building engine running while gradually reducing the fixed obligations that can trap you later.

Actionable takeaway: make the housing decision with one page of numbers

Before you make a housing change—buy, sell, refinance, upgrade, or decide to rent—write down these five numbers:

1. *True monthly cost of your current housing* (P&I + taxes + insurance + HOA + maintenance reserve).
2. *True monthly cost of the alternative* (rent, or the full cost of the new home including ownership costs).
3. *Your current monthly retirement contribution rate* (in dollars, not just a percentage).
4. *Your cash reserves after the change* (would you still have a strong emergency fund?).
5. *Your "panic threshold"* (the market drop or income disruption that would make you regret the decision).

If the new housing choice improves your life and still leaves room for reserves and retirement contributions, it can be a great decision. If it compresses your cash flow, drains your safety net, and depends on nothing going wrong, it's not just a housing upgrade—it's a bet that midlife will behave politely.

A home should be a place you live, not a monthly obligation that forces you to keep working long after you're ready to stop.

3.5 Downsizing as a Wealth Accelerator

A lot of people in their 40s and 50s live inside their retirement plan without realizing it.

It looks like a normal house: a spare bedroom that became a storage museum, a dining room used twice a year, a yard you used to love but now negotiate with like a hostile coworker. But financially, that house may be doing something quietly dramatic: it may be holding your largest pool of wealth in the least useful form.

This is the classic midlife trap: *house rich, cash poor*. Your net worth looks respectable on paper, but your monthly cash flow feels tight. Retirement contributions are "what we'll increase later." Repairs arrive like uninvited guests. And every time you think about investing more, the house absorbs the money first—because houses are extremely good at spending your cash for you.

Downsizing is the move that turns wealth that is trapped in drywall into wealth you can actually use. Not someday. Not after you sell in your 80s. Now, when compounding still has time to work and when catch-up contributions can make a meaningful difference.

The emotional myth: downsizing is defeat

Downsizing gets framed as a retreat. A smaller home means you're giving something up. In some communities, it can even feel like announcing you didn't "make it."

That story is backwards.

In midlife, downsizing can be a power move: a deliberate trade of square footage for freedom. You're not shrinking your life. You're choosing where your money lives. You're taking wealth out of an asset that produces no dividends and moving it into one that can.

It's also a way of lowering the cost of being you. If your housing costs drop, you don't just save money. You reduce the amount of income you must produce every month to stay stable. That kind of stability is what makes early retirement possible, what makes career changes survivable, and what makes layoffs less terrifying.

Downsizing isn't about living smaller. It's about making your financial life lighter.

The core math: harvested equity is only real after you subtract the bite marks

The fantasy version of downsizing is: sell big house, buy smaller house, invest the difference.

The real version includes bite marks: transaction costs, moving costs, taxes in some situations, repairs to sell, and the cost of buying into a competitive smaller-home market. If you don't subtract those, you can convince yourself you're unlocking $200,000 and then discover you unlocked $60,000 and a new set of problems.

Selling costs typically include agent commissions (if you use agents), seller concessions, and closing costs. Moving/transition costs include movers, temporary housing if there's a gap, storage, and the shockingly real "we have to buy a bunch of stuff" expenses that appear when you change spaces.

New home cash needed is the down payment and closing costs for the next place—or the cash required if you're buying outright.

That last line is the difference between downsizing that accelerates wealth and downsizing that simply rearranges your furniture.

Why this can supercharge catch-up contributions (and why timing matters)

Catch-up contributions exist for a reason: the system quietly admits that many people won't be perfectly on track by their 50s. The opportunity is real, but it requires cash flow. You can't max out retirement accounts with good intentions.

Downsizing helps in two ways:

1) It can create a one-time lump sum.

A lump sum can seed a taxable brokerage account, pay off high-interest debt that's blocking savings, or create the cash buffer that allows you to increase 401(k) contributions without fear.

2) It can lower your ongoing monthly burn rate.

If your monthly housing cost drops by $800—$2,000, you've effectively created a permanent "raise." That raise can be redirected into retirement contributions every single month.

Timing matters because you're buying *years of compounding*. A dollar invested at 45 has more time to grow than a dollar invested at 60. Downsizing at 50 instead of 62 can be the difference between "helpful" and "transformative."

A real-world case study: the ranch house that funded retirement like a second job

Marisol and Dean were 53 and 55. Their three kids were out of the house, but the family home stayed the same size as if everyone still lived there. It was a beautiful two-story in a neighborhood they loved. It was also a constant drain: landscaping, higher utilities, and repairs that arrived in waves.

They weren't struggling on paper. They had solid incomes and some retirement savings. But every year, something stole their attention and their money: the roof, the HVAC, the deck, property taxes, insurance increases. Their retirement contributions were steady but not aggressive, and they could feel that "close enough" might not be close enough.

Then a smaller house went up for sale nearby: a single-story ranch with fewer stairs, lower maintenance, and a layout that fit their life *now*, not their life fifteen years ago.

They ran the harvested equity calculation instead of guessing:

- Sold current home: $780,000
- Mortgage payoff: $220,000
- Selling costs and repairs: $55,000
- Bought ranch: $520,000
- Closing + moving + initial updates: $25,000

$$780{,}000 - 220{,}000 - 55{,}000 - 520{,}000 - 25{,}000 = -40{,}000$$

That result looks like a failure until you notice what they chose: they bought the new house mostly in cash because they wanted the payment gone. They weren't trying to harvest a huge lump sum. They were trying to harvest *monthly breathing room*.

Their old monthly housing cost (mortgage, taxes, insurance, utilities, maintenance reserve) had been about $5,200. The new one dropped to about $3,200. They freed up roughly $2,000/month.

That $2,000/month became their wealth accelerator. They increased 401(k) contributions, funded Roth IRAs when eligible, and built a taxable "retirement bridge" account. It felt like adding a second income without adding a second job.

Downsizing didn't hand them a suitcase of cash. It handed them a permanent cash-flow advantage, which is often the more powerful gift.

Downsizing that works: pick which lever you're pulling

There are two primary ways downsizing accelerates wealth, and it helps to know which one you're aiming for:

Lever A: Liquidity.

You sell, buy cheaper, and invest the difference. This is the "harvest cash" version.

Lever B: Lower fixed costs.

You may not extract a large lump sum, but you reduce the monthly cost of life (mortgage payment, utilities, maintenance). This is the "harvest monthly cash flow" version.

Both can work. Problems happen when you think you're pulling Lever A but you actually pull Lever B (or vice versa), then feel disappointed or overconfident.

If you want liquidity, you need the new home to be meaningfully cheaper *after* you account for selling and buying costs. If you want lower fixed costs, you need to focus on the total monthly ownership cost, not just the mortgage payment.

The hidden accelerant: maintenance is a tax on square footage

Maintenance costs are the least respected line item in homeownership because they feel optional until they're not.

A smaller home usually means:

- fewer systems to replace (or smaller ones)
- fewer surfaces to repair
- less exterior to paint, roof, and insure

- lower utility costs in many cases

Not always. A condo can come with high HOA dues. A smaller home in a premium neighborhood can still be expensive. But generally, the size and complexity of the home is correlated with the amount of money it will demand.

When people say, "My house is my biggest asset," they're sometimes ignoring the reality that it can also be their biggest *subscription*. Downsizing can cancel that subscription.

Insurance and taxes: smaller doesn't always mean proportionally cheaper

A smaller home can reduce costs, but don't assume every expense shrinks neatly.

Property taxes depend on assessed value and local rules, not just square footage. Insurance pricing can be driven by region-wide risk, rebuild cost inflation, and claim trends. In some areas, insurance has become a wild card: the premium can jump even if nothing about your house changed.

Two practical consequences:

- When you price a downsizing move, estimate insurance and taxes using current quotes, not last year's bill.
- Keep deductibles and emergency reserves aligned. A smaller home doesn't automatically mean a smaller deductible, and a higher deductible is only smart if you can comfortably pay it.

The point is to avoid the downsizing mirage where you move expecting costs to drop and then discover the HOA plus insurance increases ate the savings.

The alternative to selling: equity access without moving (useful, but not free)

Sometimes downsizing isn't possible or isn't desirable. Family proximity, school districts, health needs, or emotional ties can make staying the right choice.

In that case, people often look at ways to access home equity while staying put: cash-out refinancing, home equity loans or lines of credit, and (later in life) reverse mortgages.

These tools can provide liquidity, but they're not magic. They convert equity into debt, which increases fixed obligations unless handled carefully. They can be useful when the alternative is worse (high-interest debt, forced sale, inability to fund necessary expenses), but they should be evaluated with the same seriousness as any other long-term obligation.

Reverse mortgages, in particular, can be tempting because they promise cash flow in retirement years. They can also constrain flexibility later, because the loan typically comes due when the home is sold or you no longer live there. They also require that you keep paying property taxes and homeowners insurance; failing to do so can trigger default. The headline is "access equity." The fine print is "you still have responsibilities."

If the goal is to accelerate wealth in your 50s, the cleanest accelerant is usually not borrowing against the house. It's reducing the house's claim on your cash flow and investing the difference.

Geo-arbitrage: the most powerful form of downsizing (and the most personal)

If you can move to a lower cost-of-living area without sacrificing what matters most to you, geo-arbitrage can be the biggest lever available. It's downsizing plus a market shift.

This isn't just about cheaper homes. It can reduce:

- property taxes
- insurance costs (depending on hazard profile)
- everyday expenses that quietly determine your savings rate

It can also increase the non-financial costs: distance from family, changes in healthcare access, new social networks, a sense of loss. Geo-arbitrage is powerful, but it's not purely mathematical. A move that saves money but damages your support system can become expensive in other ways.

A good test is to treat geo-arbitrage like a pilot program: rent for a year in the target area if possible. Let the place prove itself before you buy.

The ``don't accidentally re-upgrade" trap

One of the most common downsizing failures is psychological: you sell a big house, then buy a smaller one that is so upgraded—premium finishes, top-tier neighborhood, every bell and whistle—that the monthly cost doesn't drop much. You traded space for status and called it downsizing.

There's nothing wrong with buying something beautiful. The danger is thinking you made a wealth move when you made a lifestyle move.

If your goal is to accelerate retirement savings, you need at least one of these to happen:

- investable cash created at the move, or
- a meaningful reduction in monthly housing costs.

If neither happens, the move may still be worth it, but it won't fund your catch-up plan.

Actionable takeaway: run the downsizing ``profit and cash-flow" test before you list

Do this on one sheet of paper (or one spreadsheet tab) before you fall in love with Zillow photos:

1. Estimate your sale price conservatively and subtract mortgage payoff.
2. Subtract selling costs, repairs, and moving/transition costs.
3. Estimate the all-in monthly cost of the new housing (including HOA, taxes, insurance, and a maintenance reserve).
4. Calculate two outputs:
    - *Investable cash created* (if any).

- *Monthly cash flow freed up* (if any).

5. Assign the freed-up money a job *in advance*: automatic retirement contributions, a taxable brokerage transfer, or debt payoff. If it doesn't get assigned, it will get consumed.

The win isn't a smaller house. The win is money that stops hiding in your walls and starts working in your accounts, month after month, until your future stops feeling like a deadline and starts feeling like an option.

3.6 Estate Planning Essentials

Estate planning has a branding problem.

It sounds like something for people with yachts, complicated family trees, and a dedicated room for antiques. If you're 45 with a mortgage, a 401(k), and a phone full of school pickup texts, it can feel irrelevant—or worse, like inviting bad luck by putting it on the calendar.

But estate planning isn't about predicting death. It's about preventing bureaucratic chaos. It's the paperwork that makes your financial life transferable instead of tangled. And in midlife, when you've finally built real assets and real responsibilities, the cost of leaving things "for later" is usually paid by someone you love.

There are three documents that do most of the heavy lifting for ordinary families. You don't need to memorize Latin phrases or build a trust empire to get enormous benefit. You need the *Big Three*:

- a will
- powers of attorney (financial and medical)
- beneficiary designations (and related account titling)

If that list feels underwhelming, good. Underwhelming is exactly what you want from a plan that's supposed to work on the worst day.

The real question: who runs your life if you can't?

When people hear "estate planning," they think "who gets my stuff when I die." That's only part of it, and for midlife families it's often not even the most urgent part.

The more immediate question is: *Who can make decisions and pay bills if you're alive but unable to act?*

A stroke. A car accident. A severe illness. Even a short hospitalization can put your household in a weird limbo where someone needs legal authority to handle routine tasks: accessing accounts, managing insurance claims, signing documents, dealing with your employer, keeping the mortgage current.

Without the right documents, your spouse or partner may discover an infuriating truth: love is not legal authority.

The goal of the Big Three is to make sure your household can keep functioning without needing a court's permission.

The will: not a wealth document, a traffic cop

A will is a set of instructions for what happens to assets that are subject to probate—the court-supervised process that validates a will and oversees distribution. Probate is not always a disaster, but it's rarely fast, and it can be more expensive and stressful than people expect.

A will does a few core jobs:

- names who receives certain assets *that don't have a beneficiary designation*
- names an executor (the person who handles your estate administration)
- if you have minor children, names a guardian (arguably the most emotionally important line on the page)

The guardian nomination is why even families with very little money still need a will. If you have children under 18, "we don't have much" is not a reason to skip it. In a crisis, you want your voice to exist on paper.

A will is also where many people accidentally overestimate what it can do. A will is not a magic override. It doesn't necessarily control everything you own, and it can lose in a fight with other paperwork.

Which brings us to the most important and most commonly botched part of estate planning.

Beneficiary designations: the forms that often outrank your will

Many of your biggest accounts transfer by contract, not by will. Retirement accounts (like 401(k)s and IRAs), many life insurance policies, and sometimes brokerage accounts with a Transfer on Death (TOD) registration pass to the named beneficiary automatically. That transfer can happen outside of probate.

This is convenient. It's also dangerous when your forms are outdated.

Here's the rule that saves families from expensive mistakes:

> If an account has a beneficiary designation, that designation often controls who gets the money, even if your will says something different.

This is not a niche technicality. It's one of the most common ways well-meaning plans fail. People spend time drafting a thoughtful will and then accidentally leave their 401(k) to an ex-spouse because the beneficiary form hasn't been updated since the Obama administration.

If you do nothing else after reading this, log into your retirement accounts and verify the beneficiaries. It is boring work with an unusually high return.

A real-world case study: the ``we wrote a will'' family who still got burned

Kevin and Rina were in their early 50s. Second marriage. Each had adult children from prior relationships, and they shared a home and a set of financial goals that made sense to them: if one died, the survivor should have enough stability to stay in the home and not panic. Eventually, whatever was left would go to their combined kids.

They did a will online, felt responsible, and moved on.

When Kevin died unexpectedly, Rina assumed the will would guide everything. It did not. Kevin's largest asset besides the house was his 401(k), and the beneficiary designation still named his first spouse. No one had thought to change it because, in their minds, the will was the update.

The outcome wasn't just financial. It was emotional shrapnel. The kids were confused and angry. Rina was stunned. The will didn't matter for that account. The beneficiary form did.

Could it have been fixed? Often, yes—with one login, one form, one follow-through. But the fix didn't happen while Kevin was alive, so everyone learned the lesson the hardest way.

Estate planning isn't just legal documents. It s *coordination.*

Account titling: how you own something can be as important as who you want to get it

Beneficiary forms are only part of the non-probate universe. Account titling—the legal way an asset is owned—also affects what happens at death and during incapacity.

Common ownership structures include:

- *Individual ownership*: one person owns it; it may go through probate unless there's a beneficiary/TOD.
- *Joint ownership with right of survivorship*: the surviving owner typically becomes the full owner automatically.
- *Tenants in common*: each owner has a share that can pass according to their will (often involves probate).

The details vary by state and account type, but the principle is universal: your estate plan is not only what your documents say; it's how your assets are wired to transfer.

A common midlife mistake is accidentally creating complexity by scattering assets across different titling styles without realizing it. Another is assuming "my spouse will handle it" while maintaining individual accounts with no beneficiary designations and no authority documents. That's how you end up with accounts frozen at the exact moment liquidity matters.

Powers of attorney: the documents you hope you never need and should absolutely have

A power of attorney (POA) is a document that allows someone else (your agent) to act on your behalf.

There are two broad types most families need:

Financial power of attorney.
This authorizes someone to handle money matters: paying bills, managing accounts, signing documents, dealing with property. Many versions are "durable," meaning they remain effective if you become incapacitated. Without it, your spouse may be forced into a court process to gain authority to manage routine financial tasks.

Medical power of attorney (health care proxy).
This authorizes someone to make medical decisions if you cannot. It often works alongside documents like a living will or advance directive, where you express your preferences about end-of-life care.

The reason these documents matter in midlife is that incapacity planning is not rare, and it's not always permanent. People can be temporarily unable to act due to surgery, complications, mental health crises, or medication effects. Life doesn't politely wait for you to recover before it sends the mortgage statement.

A good POA setup is like a spare key you never want to use. The day you need it, you need it immediately.

The household operations problem: someone needs a map

Even with perfect legal documents, families often struggle because the surviving spouse can't find information. The accounts are online, protected by two-factor authentication that goes to a phone that is now locked. Bills are on autopay, but no one knows which account pays what. Insurance policies exist, but where?

The fix isn't complicated. It's operational.

Create a simple "financial map" that includes:

- a list of accounts (banking, retirement, brokerage) and where they are held
- insurance policies (life, home, auto, disability) and the insurer contact information
- recurring bills and how they're paid
- key professionals (CPA, attorney, financial planner) if you have them
- where the estate documents are stored (and how to access them)

You don't have to include passwords in a document that lives in a drawer. But you do need a plan for access. Many people use a password manager with an emergency access feature. Some keep instructions in a secure location. The method matters less than the reality: if the household can't access accounts, the plan is theoretical.

A quick word about taxes: most families aren't fighting the estate tax, they're fighting confusion

Estate tax headlines are excellent at stealing attention from what matters.

Most midlife families will never owe federal estate tax. The more common (and more expensive) problem is a mismatch between intentions and paperwork: outdated beneficiaries, unclear ownership, missing POAs, and documents that don't reflect the current family structure.

If you do have significant wealth, business interests, or complicated family dynamics, professional advice is worth it. But even then, the foundational work is the same: clear documents, consistent beneficiary designations, and coordinated titling.

The biggest estate planning win for most readers is not a fancy trust structure. It's preventing the state from playing "guess what they meant."

Funeral costs: grief is not the moment to negotiate

A death in the family creates a financial moment that is both urgent and emotionally charged. People make expensive decisions quickly because they want to honor someone, avoid conflict, or simply end the painful logistics.

The practical antidote is gentle planning: a conversation about preferences and a written note about what matters. Even a small earmarked fund can prevent a credit card from becoming the default funeral financing plan.

One consumer tip that matters: funeral homes must provide itemized price information and allow you to choose only the goods and services you want. Having that knowledge in advance can protect a family from making a five-figure decision while in shock.

The ``blended family" and ``second marriage" alert

If you're remarried, have children from prior relationships, or support family members outside the current household, estate planning becomes less optional and more essential.

Blended families are where default rules (and outdated beneficiary forms) create the most unintended outcomes. Many couples want a plan that provides stability for the surviving spouse but ultimately preserves an inheritance for children from a prior marriage. That goal is achievable, but it requires careful coordination—and often professional guidance—because the simple default of "everything to spouse" can conflict with "eventually to kids."

If your family situation is at all complex, the highest-leverage move is to get your beneficiary designations aligned with your actual intention and to make sure your will and powers of attorney match your current life.

Actionable takeaway: do the Big Three audit in one weekend

Pick a Saturday morning and do an estate planning audit that is more operational than philosophical:

1. *Beneficiary check*: Log into every retirement account and life insurance policy you can find. Confirm beneficiaries and contingent beneficiaries. Screenshot confirmation pages or save PDFs where available.
2. *Document check*: Locate your will and your financial/medical powers of attorney. If you can't find them in five minutes, assume your family won't find them in five hours.
3. *Alignment check*: Make sure the names on the documents match reality (marriage, divorce, deaths, address changes). Confirm the executor and agents are still the right people.
4. *Map check*: Create a one-page list of accounts, policies, and where documents are stored. Tell your spouse (and a trusted person) where to find it.

A good estate plan doesn't feel inspiring. It feels slightly boring and deeply relieving. It's the quiet confidence that if something happens, your family won't have to fight paperwork while they're trying to take care of each other.

Chapter 4

Investing for Growth: The Catch-Up Engine

If you are starting late, you might feel like the magic of compound interest is a party you missed in your twenties. But the engine hasn't stalled; it just needs higher octane fuel. This chapter moves beyond simple saving to building a portfolio that works as hard as you do, proving that even with a shorter runway, smart, aggressive, and efficient investing can still help you take flight.

4.1 Compounding on a Deadline

At 42, you don't need a lecture about time. Time is already loud. It's in the birthday candles that suddenly look like a small bonfire, the parents who now need help instead of giving it, the kid who somehow has a driver's permit. Time is also in your retirement account balance, sitting there like a number that refuses to match your effort.

And when people talk about investing, they tend to talk the way people talk about learning a language: start early or don't bother. You hear, "Compounding is magic," and it sounds like a magic trick you missed because you showed up after intermission.

Here's the thing: compounding isn't a reward for perfect timing. It's a *mechanism*. It works whenever you feed it. The difference between starting at 25 and starting at 45 is real, but the story most people tell about that difference is incomplete. They focus on the calendar and ignore the throttle.

The throttle is contributions.

Compounding is what happens when growth starts growing too. But in the real world—where paychecks are lumpy, careers have plot twists, and life gets expensive at exactly the wrong times—compounding rarely begins with a huge pile of money. It begins with a habit. The "engine" needs fuel. For late starters, more fuel can make up for less runway.

The myth: "If I didn't start early, I'm too late"

The most damaging misunderstanding about compounding is that it needs forty years to matter. That's like saying strength training only works if you start in high school. It's true that earlier is easier. It's also true that humans get strong in their forties all the time.

Compounding is simply repeated multiplication. If your investments earn a return, and you leave those returns in place, your base gets larger. Then the same percentage return applies to a larger base. That's the whole trick.

What changes when you start later is not whether compounding works. What changes is that *your contributions do more of the heavy lifting early on*. When you're 25 with decades ahead, it's easy to let the market do most of the work. When you're 45, you're asking the market to sprint while also asking your money to grow legs. That's when contributions stop being a side character and become the lead.

A simple way to see it: if you invest for 20 years, you don't get many "doublings." Even at a strong long-term return, you might get two-ish doubling periods, maybe fewer. That means the dollars you add along the way matter enormously. Increase the dollars, and you increase what gets the chance to compound.

Catch-up compounding: how higher contributions impersonate time

Imagine two people who both want to retire at 67.

Early Starter begins investing at 25. *Late Starter* begins at 45.

If Early Starter saves a little and Late Starter saves a lot, Late Starter can narrow the gap dramatically. Not with heroics, not with lottery-ticket stocks, and not by "taking crazy risk." Just by contributing at a level that makes the math work.

Here's the core idea in plain English:

> Time multiplies money. Contributions *add* money.
> When time is limited, you lean harder on adding.

This isn't motivational fluff; it's the structure of the formula. The future value of a growing investment is driven by:

- what you start with,
- what you add,
- how long it has to grow,

- and the return you earn.

If one of those inputs (time) is smaller, you can still increase another input (contributions). You can't rewind the clock, but you can widen the channel.

A useful mental model: when you start later, the portfolio behaves less like a snowball rolling down a hill and more like a flywheel you're pushing into motion. In the early years, most of the progress you see comes from what you put in. Over time, if you keep pushing, the flywheel starts carrying itself. The "compound" part eventually takes over more of the workload, but only if you build the base first.

A real-world case: "Derek, 46, finally gets serious"

Derek (not his real name) was a solid earner who did what many responsible adults do: he built a life. Mortgage, childcare, a period of helping his mom with medical bills, and a few career detours that were good for his sanity but not great for his 401(k). At 46, he had about $85,000 in retirement savings. He wasn't broke. He wasn't reckless. He was just underfunded for the timeline he wanted.

What changed wasn't a sudden obsession with finance. It was one uncomfortable afternoon with a spreadsheet and a simple realization: if he kept contributing $300 a month because it "felt responsible," he would arrive at retirement with a number that looked fine until you translated it into monthly income. The future version of Derek wasn't going to be impressed by the good intentions of the past version.

So Derek did three practical things:

1. He increased his 401(k) contribution to a level that made him wince a little (because painless changes are usually too small).
2. He automated an annual increase timed to his raise season, so the contribution percentage climbed without repeated willpower.
3. He stopped treating investing like a mood. The contributions happened whether markets were up, down, or dramatic.

Derek didn't become a day trader. He didn't "outsmart" anyone. He just got the fuel line right. Over the first few years, most of his progress came from contributions. Then the account balance got big enough that market growth began to matter more. That handoff—from "I'm building the base" to "the base is building itself"—is the moment late starters need to manufacture.

Contribution rate beats calendar time more often than people expect

People love the romance of the early start, but they underestimate how much money can be built in 15–22 years when contributions are high and consistent.

To make that concrete, compare two simplified paths to age 67:

- *Early Starter* (*Low Savings*): starts at 25 with modest monthly contributions.
- *Late Starter* (*High Savings*): starts at 45 with significantly higher monthly contributions.

Assume the same long-term average return for both. (Real life is messier, but the mechanism is the point.) The early starter has more time, but the late starter has a steeper savings rate. The late starter's curve can look almost vertical in the final stretch—not because compounding suddenly becomes magic, but because the base is finally big and the contributions have been relentless.

When readers see this drawn, something clicks: the late starter isn't "starting behind" forever. They're just operating a different machine. Early Starter uses time as the main lever. Late Starter uses a higher savings rate as the lever.

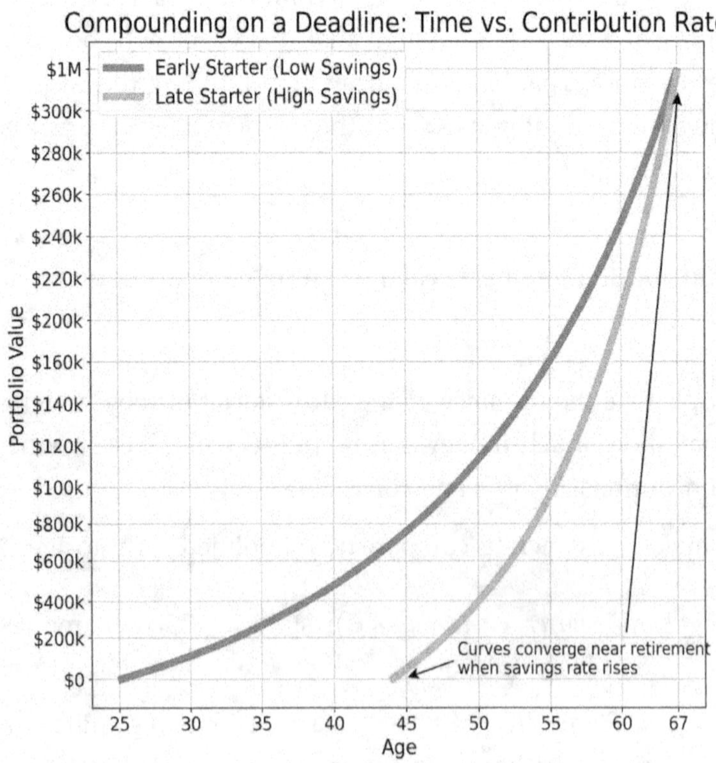

Why late starters often underestimate what they can do

A late start creates a specific kind of mental trap: you look at the gap and assume the only way to close it is with a *higher return.* That's the seductive fantasy—the "if I just pick the right fund" storyline. It's also the storyline that sells expensive products and encourages reckless bets.

But for most households, the most controllable lever isn't return. It's *how much of your income you keep and invest*. That lever doesn't require predicting markets. It requires designing a life that funds the future.

There's also a quieter psychological issue: many people treat their current contribution level as if it's a personality trait. "I'm a 6% contributor." As if their retirement plan came with their eye color. Contribution rates are not identities. They're settings.

And settings can be changed.

The "first $100,000" problem (and why it feels slow)

Late starters often quit right before it gets good, because the early phase feels insultingly slow. You do the responsible thing. You set up the contributions. You watch your account. Then the market has a dull year, or a bad year, and your balance looks like it's barely moving.

That's normal. Early on, the account is small enough that even a strong market year doesn't add much in dollar terms. A 10% gain on $20,000 is $2,000. Nice, but it won't feel like a life change. This is where consistency matters more than enthusiasm.

The good news is that the early phase is also when contributions have the most power to change your trajectory. If you're 45 and you can increase your annual contributions by $6,000, you didn't just add $6,000. You added $6,000 *plus the chance for every future year of returns on that $6,000*. That's compounding's real job: turning today's sacrifice into tomorrow's leverage.

A deadline changes the strategy, not the possibility

A deadline doesn't mean giving up on growth. It means getting sharper about the inputs you control.

Three practical realities come with a shorter runway:

1. *You can't wait for motivation.* The plan has to run on autopilot.
2. *You can't afford chronic under-contribution.* Saving "something" is not the same as saving "enough."
3. *You need a plan that survives bad years.* Markets will have them. Your plan has to assume that and keep going.

The deadline also creates a strange advantage: clarity. When retirement is far away, it's easy to treat investing like a hobby. When it's closer, investing becomes a system. Systems are powerful because they don't require you to be in the right mood.

What "higher contributions" actually looks like in real life

The phrase "save more" is useless advice unless it becomes a specific redesign of cash flow. Late starters who succeed tend to use some combination of these moves:

- *Raise the floor before you raise the ceiling.* Start with a contribution you can sustain in a boring year, then increase it deliberately.
- *Use pay raises as a stealth funding source.* If your paycheck increases, divert part of the increase before lifestyle expands to consume it.
- *Treat debt payoff as future contribution capacity.* The day a car loan ends should be the day an automatic investment begins.
- *Separate "can't" from "haven't tried."* Many households assume a higher contribution is impossible without first testing what a tighter month actually feels like.

Notice what's missing: a requirement to be extreme. This is less about living on rice and more about reclaiming dollars that leak away unnoticed—subscriptions you don't use, "small" purchases that aren't small when repeated, and the quiet creep of lifestyle inflation.

Late starters also tend to benefit from a simple truth: you don't have to do it perfectly forever. You have to do it consistently for long enough that the flywheel starts spinning. After that, the growth becomes more self-sustaining, and the required contribution rate can feel less like a punishment and more like a normal line item.

A brief, honest word about assumptions

Charts like the one above are not promises. Markets don't deliver the same return every year, and retirement planning is vulnerable to the sequence of returns (whether bad years happen early or late). But the chart captures something sturdier than a forecast: the idea that contributions change the geometry of the problem.

When the timeline is short, you want your plan to be resilient even if returns are lower than hoped. The most reliable way to build that resilience is to increase the amount you're putting in and to keep costs and taxes from silently draining the compounding process. You don't need every year to be a winner; you need a plan that *keeps showing up*.

The takeaway that actually moves the needle

Pick a retirement contribution level that feels slightly too high, then make it automatic and give it a ramp.

Here's a practical way to do that without relying on a personality transplant: set your contribution so that it's uncomfortable but not destabilizing, then schedule an automatic increase of 1% of pay every year (or every time you get a raise) until you reach the level your future self needs. If you're eligible for catch-up contributions as you approach your fifties, treat them not as an optional bonus, but as the default setting of the plan.

Compounding doesn't require a perfect past. It requires a funded present.

4.2 Risk, Return, and the Fear of Loss

The most common late-starter investing plan is not a plan. It's a reflex.

A headline hits. The market drops. Your stomach drops with it. You picture the money you can't afford to lose—college bills, a medical curveball, the version of you who would like to stop working before your knees file a complaint. And you do the one thing that feels like control: you move to cash.

Cash has a special talent. It can make you feel safe while quietly putting your future on layaway.

That's not a moral failing. It's wiring. When people say they're "risk-averse," they usually mean they're "loss-averse": the pain of losing $10,000 feels sharper than the pleasure of gaining $10,000. Your brain doesn't do symmetry. It does survival. In the short term, that instinct is brilliant. In the long term—especially when retirement is within sight—it can become a trap disguised as prudence.

The fix is not to become fearless. The fix is to redefine what *risk* actually is.

Risk isn't just a drop on a chart

The investing world loves a particular definition of risk: volatility. Volatility means prices move around, sometimes violently. It's the market's mood swings made visible.

Volatility is real. It can be brutal. But it's only one kind of risk, and it's the one that gets all the attention because it's loud. You can see it in red numbers. You can feel it in your chest at 2 a.m. when you're scrolling financial news like it's a crime report.

There's a quieter risk that doesn't spike your pulse because it doesn't arrive as an emergency. It arrives as a slow leak.

That risk is purchasing power.

Purchasing power is what your money can actually *buy*. If your savings earns 1% and prices rise 4%, you didn't "stay even." You lost ground. You just lost it without a dramatic headline. Inflation is not a crash; it's a tide. If you stand still in the water long enough, you end up somewhere you didn't choose.

So risk has at least two faces:

- *Volatility risk*: the risk your account balance drops when markets fall.
- *Shortfall risk*: the risk your money doesn't grow enough to fund your life, especially after inflation.

For many people over 40, the bigger danger is not that their portfolio will swing. The bigger danger is that it won't grow.

The inflation shark

If volatility is the thunderstorm, inflation is the shark under the surface. You don't always see it, but it's moving, and it never gets tired.

Here's what makes inflation so sneaky: it turns "conservative" into "expensive." You can do everything that looks responsible—avoid debt, keep savings liquid, refuse to gamble—and still end up taking the biggest gamble of all: betting that your future expenses won't be that high.

Imagine a couple, both 52, who want to retire at 67. They don't want anything extravagant. A paid-off house, some travel, helping the grandkids, not panicking about every grocery trip. They estimate they'll need $70,000 per year to live comfortably in today's dollars.

If inflation averages 3% a year, $70,000 today is roughly $109,000 in 15 years. Same lifestyle, bigger number. (That's not a doomsday scenario; that's just arithmetic.)

Now picture them sitting on a large cash position because "cash is safe." Safe from what? Safe from volatility, yes. Safe from inflation, no. Safe from running out of money, not necessarily. They may avoid watching their balance dip, but they could still arrive at retirement with a portfolio that can't buy their retirement.

The frightening part is that this failure feels polite. Nothing explodes. No one calls to warn you. You just wake up one day and realize your future has gotten more expensive while your plan stayed the same price.

A real-world case: the "2022 whiplash" lesson

Consider what happened to a lot of careful people in 2022. Markets fell. Bonds, which many people assume are always the calm part of the portfolio, also had a rough year as interest rates rose. It was a gut-check year for anyone who had started to think of investing as a smooth escalator.

Plenty of people reacted in a very human way: they sold investments after the declines and moved to cash or money market funds, telling themselves they'd "get back in when things look better."

Some did. Many didn't. The market doesn't send a calendar invitation for the "all clear." It recovers in the same messy way it falls: suddenly, unevenly, and often while the news still sounds grim.

Now layer in inflation. 2022 wasn't just a down market year; it was also a year where prices were rising rapidly. So the person who sold after losses didn't merely lock in the drop. They also parked in an asset that, after inflation, was shrinking in real terms.

That double hit is what late starters can't afford: losing growth years *and* losing purchasing power years.

The takeaway isn't that 2022 was "abnormal." The takeaway is that markets periodically create moments that force you to choose which risk you're going to accept. You don't get to avoid risk; you get to pick the kind you'll live with.

The hidden math of "feeling safe"

Feelings are data. They just aren't always directions.

When someone says, "I'm going to cash until things calm down," they are usually trying to avoid regret. If you stay invested and the market drops more, you feel foolish. If you move to cash and the market rebounds, you feel foolish. Humans hate being the person who made the "wrong" move, so we reach for what feels defensible.

Cash feels defensible because it has a stable number. Your statement doesn't insult you. It doesn't wake you up with a surprise. It gives you the emotional experience of safety.

But the question that matters is not "Will I feel safe this month?" The question is "Will I be safe twenty years from now?"

This is where the concept of *real return* matters. Real return is your investment return after inflation. A 5% return in a 2% inflation world is roughly a 3% real return. A 2% return in a 4% inflation world is roughly a *negative* 2% real return. Your money is growing in digits and shrinking in usefulness.

Late starters often need their investments to do real work. Not headline work. Real work.

Your brain is built for the wrong time horizon

If you find yourself checking your account balance constantly, you're not being responsible. You're being recruited into short-term thinking.

Frequent checking turns a long-term plan into a series of daily verdicts. And daily verdicts make even a good portfolio look like a bad idea. The market goes down far more often in short windows than in long ones. When you evaluate your portfolio like a one-year project, you start behaving like you have a one-year time horizon—even if you don't.

That's myopic loss aversion in action: the more often you look, the more often you see losses, and the more your brain concludes that the world is unsafe. It's not weakness; it's a predictable response to repeated negative stimuli.

The practical fix is not denial. It's reducing unnecessary triggers.

- Stop checking balances like social media.
- Pick a schedule (monthly, quarterly) and stick to it.
- If you must check, track contributions and savings rate, not daily market moves.

You can't out-discipline your nervous system forever. You can, however, design your environment so you aren't poking the bruise every morning.

The adult version of risk tolerance

"Risk tolerance" sounds like a personality quiz. In reality, it's more like a two-part job interview.

Part one: *Can you emotionally handle your portfolio dropping without making a permanent mistake?*

Part two: *Can your plan financially survive that drop without forcing you to sell?*

Those two questions matter more than any label like "aggressive" or "conservative." A portfolio that is mathematically perfect but emotionally unbearable is not a solution. It's a future panic sale waiting for a catalyst.

For late starters, the goal is not to chase the highest return. The goal is to take *enough* growth exposure to beat inflation and build wealth, while also building enough stability that you don't abandon the plan at the worst possible moment.

That stability often comes from something that isn't talked about enough because it isn't flashy: a buffer.

The buffer that makes courage unnecessary

The reason market drops feel existential is that, for many households, they *are*. If you don't have an emergency fund, a job loss or medical bill colliding with a down market can force you to sell investments at the exact wrong time.

So the most practical way to become a better investor isn't learning more ticker symbols. It's building a cash buffer that protects you from needing your investments on a bad day.

Think of cash not as an investment competing with your portfolio, but as an insurance policy protecting your portfolio.

A simple framework:

- *Cash for emergencies*: money you can access quickly without caring what the market is doing.
- *Investments for retirement*: money that needs time and growth to do its job.

When those roles are mixed, everything feels risky. When they're separated, you stop asking your retirement account to also function as a checking account.

People who have a buffer don't need to be brave. They just need to be consistent.

"But what if I invest now and the market crashes?"

This is the question that keeps people stuck. It sounds like a request for a prediction, but it's actually a request for emotional safety.

Here's the honest answer: yes, the market could drop after you invest. It could drop a lot. It will drop at some point; that's not pessimism, that's the admission price for growth.

The more helpful question is: *What's my plan if it drops?*

If your plan is "sell to avoid losing more," then you're not investing; you're speculating with a panic button. You'll buy high (when you feel safe) and sell low (when you feel unsafe), which is how smart people accidentally manufacture terrible returns.

A better plan has two parts:

- *A contribution schedule you can keep during a downturn.*
- *A portfolio mix that you won't abandon when it gets ugly.*

You can't control the storm. You can control whether you run out into it.

A concrete example: two savers, one "safe," one safer

Take two hypothetical savers, both 50, both aiming for 67, both with $200,000 invested today and both adding $1,000 per month.

Saver A keeps everything in cash-like instruments earning 3% annually. Saver B invests in a diversified portfolio expected to earn 7% annually over the long run, with plenty of volatility along the way.

If inflation averages 3%, Saver A's real return is close to zero. Saver B's real return is closer to 4%. Over 17 years, that difference isn't just a little extra. It can be the difference between "retirement is a choice" and "retirement is a negotiation."

Saver A will sleep better during market drops, but Saver A may sleep worse at 66. Saver B will have some uncomfortable years, but Saver B has a better shot at purchasing power—the ability to fund a real life, not just maintain a number on a screen.

This is the redefinition of risk that matters after 40: the risk is not that your balance fluctuates. The risk is that your future becomes unaffordable.

How to stop treating volatility like a personal attack

Volatility feels insulting because it suggests you did something wrong. You invested. The market fell. Your brain wants a culprit, and the easiest culprit is you.

But markets don't respond to your virtue. They respond to millions of buyers and sellers reacting to growth, inflation, rates, geopolitics, corporate earnings, and fear itself. A down year isn't a sign you "messed up." It's a normal season in a system built to move.

One mindset shift helps more than any chart: stop thinking of a market drop as "I lost money," and start thinking of it as "prices changed."

If you're still contributing, lower prices mean your new contributions buy more shares. That doesn't make downturns fun, but it turns them from purely destructive events into mixed events: painful on statements, useful for buyers.

The market punishes impatience and rewards persistence, not because it's moral, but because that's how compounding interacts with human behavior. The investor who can keep contributing through ugliness has an advantage over the investor who needs good feelings to participate.

The action that matters most: define your "panic rules" now

When people panic, they don't rise to the occasion. They fall to their defaults. So set the default.

Write a one-paragraph "panic rule" while you're calm. Keep it simple enough that you'll actually follow it:

- Under what conditions, if any, will you change your investments?
- What will you *not* do (for example, "I will not sell stocks because of headlines")?
- Who will you talk to before making a move (a spouse, a planner, a trusted friend)?

Then build the support structure that makes those rules realistic: an emergency fund, automated contributions, and a portfolio whose ups and downs you can survive both financially and emotionally.

That's what "risk management" looks like in real life: not eliminating fear, but preventing fear from being in charge of the keyboard.

4.3 Asset Classes Decoded

Walk into any workplace retirement plan website and you'll be greeted by a wall of fund names that sound like they were generated by a corporate thesaurus: "Strategic," "Opportunities," "Dynamic," "Enhanced," "Absolute." The words are supposed to make you feel like something intelligent is happening.

But under the hood, most portfolios are made from three plain materials. Everything else is packaging.

Stocks are the growth engine. Bonds are the stabilizers. Cash is the emergency padding. Get clear on what each one is *for*, and you stop shopping for investments the way people shop for skincare—buying whatever promises miracles and hoping it doesn't burn.

A useful analogy is a car. Not because investing is a road trip, but because the parts have jobs that match what your money needs to do after 40.

- *Stocks are the gas pedal*: they're how you close the gap.
- *Bonds are the brakes and suspension*: they're how you stay on the road when things get rough.
- *Cash is the airbag*: it doesn't help you win the race, but it can save you in a crash.

That's the whole system. Once you understand those jobs, fund names stop being persuasive, and your decisions get calmer.

Stocks: buying slices of human ambition

When you buy a stock, you're buying ownership in a company. Not a prediction about a ticker symbol, not a bet on a headline—a claim on the future profits of a real business. If the business grows over time, the value of your slice tends to grow too.

Stocks feel risky because the market reprices them constantly. Your ownership doesn't change every day, but the price does. That's volatility: the market's opinion, updated in real time, sometimes based on fundamentals and sometimes based on pure mood.

Over long periods, stocks have historically been the main source of growth that beats inflation by a meaningful margin. That matters for late starters because inflation is a quiet thief and retirement is an expensive promise. If you need your money to grow into something larger than the sum of your contributions, stocks are usually doing the heavy lifting.

The catch is emotional. Stocks can make you feel brilliant and stupid in alternating months. They are the gas pedal, but gas pedals don't come with comfort.

A practical way to think about stock risk: it's not that stocks are "bad" or "unsafe." It's that stocks are *uncertain on your schedule*. They can be down for a year, two years, or longer. The reward for tolerating that uncertainty is the possibility of real growth.

How to own stocks without turning your retirement into a personality test

Many people hear "stocks" and picture picking individual companies—the way you might pick a restaurant for dinner. That's not required, and for most busy adults, it's not even desirable.

A simple approach is to own stocks through pooled funds, usually mutual funds or exchange-traded funds (ETFs). These are baskets that hold many stocks. Instead of betting on one company, you can buy a tiny slice of hundreds or thousands. This is how you turn "stock investing" from a high-stress hobby into a background system.

Index funds are a specific type of pooled fund that simply aims to match a market index (like a broad U.S. stock index) rather than trying to outsmart it. The advantage isn't glamour. The advantage is that it keeps your hand off the "do something!" lever that investors tend to pull at exactly the wrong time.

If stocks are your growth engine, broad stock funds are how you get that engine without needing to be a mechanic.

Bonds: the part of the portfolio that behaves like an adult

Bonds are often described as "safe," which is both true and misleading.

A bond is essentially a loan you make to a borrower—a government, a city, or a corporation. In exchange, you get interest payments, and you expect to get your principal back at maturity (when the bond comes due). That steady interest is why bonds have a reputation for stability.

But bonds have their own risks, and ignoring them is how people get blindsided while believing they were being cautious.

The biggest bond risk most people meet in real life is interest-rate risk. Bond prices generally move inversely to interest rates: when rates rise, existing bonds with lower interest rates become less attractive, so their prices tend to fall. When rates fall, older bonds with higher rates become more attractive, so their prices tend to rise.

This is why bonds can have bad years, sometimes surprisingly bad ones. It's not because bonds are "broken." It's because bond math is doing what bond math does.

Duration is a key concept here. Duration is a measure of how sensitive a bond (or bond fund) is to interest-rate changes. Longer duration generally means more sensitivity: bigger price drops when rates rise, bigger price gains when rates fall. You don't need to become a bond trader, but you do want to understand that "bond" is not a synonym for "cash."

So what job do bonds do?

They reduce the chance that your portfolio experiences a catastrophic drop at the exact wrong time. They also give you a steadier source of return and, in many market environments, they cushion stock declines. Not always, but often enough to matter.

In the car analogy, bonds are the suspension and brakes. They don't make the car fast. They make it controllable.

Bonds aren't one thing: what you're actually choosing

Saying "I have bonds" is like saying "I have food." It matters *what kind*.

A few broad categories you'll run into:

- *U.S. Treasury bonds*: backed by the U.S. government; generally considered among the safest in terms of credit risk (the risk of default).
- *Investment-grade corporate bonds*: loans to financially strong companies; usually pay higher interest than Treasuries because they carry more credit risk.
- *High-yield ("junk") bonds*: loans to riskier companies; higher interest, higher risk; can behave more like stocks during stress.
- *Municipal bonds*: issued by states and cities; may have tax advantages for some investors.

Bond funds bundle many bonds together, which can reduce the impact of any single borrower getting into trouble. But bond funds don't "mature" the way an individual bond does, and their prices can move day to day. Again: bonds can be stabilizing without being perfectly stable.

If you're investing primarily through retirement accounts and you want simplicity, many people use broad, high-quality bond index funds as their main bond exposure. The goal isn't to squeeze every last basis point. The goal is reliability.

Cash: safety with an expiration date

Cash gets disrespected in investing conversations because it's not exciting. That's exactly why it's useful.

Cash is money you can access quickly with minimal risk of a sudden drop in nominal value. Think: savings accounts, money market funds, short-term Treasury bills. Cash is what you reach for when life happens.

And life happens with creative timing.

The problem is that cash is a poor long-term growth tool. After inflation and taxes, it often delivers little to no real return. That means cash has a critical job, but it's a job with boundaries.

Cash is not the growth engine. Cash is the airbag.

An airbag is brilliant engineering, but you don't drive around *on* the airbag. You keep it there for the moment you need it, and you hope you never do.

When people keep too much in cash for too long, it's usually because they're trying to eliminate anxiety. The trouble is that anxiety is not the same as danger. A portfolio designed to eliminate anxiety can accidentally increase danger by failing to grow.

A better way to use cash is strategic: enough to keep you from selling investments in a crisis, not so much that it quietly starves your long-term plan.

A special kind of cash: inflation-linked savings (I Bonds)

Some savers want cash-like stability but hate the feeling of inflation eating their money alive. One tool the U.S. offers is Series I Savings Bonds (often called I Bonds). Their interest rate is tied to inflation through a composite rate that resets every six months. Interest accrues monthly and compounds semiannually.

I Bonds have rules: purchase limits, holding period requirements, and potential penalties for redeeming too early. They're not a magic account you can use like a checking account. But they can be useful as part of the "airbag" system for people who want a place to keep some savings that's designed to keep pace with inflation.

The bigger point isn't that everyone needs I Bonds. The point is that "cash" isn't only one option, and your cash choices should be made with inflation in mind, not just comfort.

Case study: Maria's portfolio problem wasn't performance—it was purpose

Maria, 49, had done what a lot of smart, capable people do: she accumulated financial products instead of building a portfolio.

She had:

- a 401(k) invested in a "balanced" fund she picked years ago,
- a brokerage account with a handful of individual stocks she bought because she liked the companies,
- a savings account that had grown large because it felt good to see a stable number.

Maria wasn't careless. She was scattered. Every account had a different logic, and none of them clearly matched a job.

The moment that changed things was not a market event. It was a personal one: her company announced layoffs. Maria kept her job, but the experience exposed the weak link in her setup. She realized her "investments" were doing three different jobs poorly because she hadn't assigned roles.

She rebuilt using the car-part mindset:

- *Airbag* (*cash*): she set a specific emergency fund target based on her household's needs and job stability.
- *Suspension* (*bonds*): she held high-quality bonds inside her retirement accounts to reduce portfolio whiplash.
- *Engine* (*stocks*): she consolidated the stock exposure into broad funds rather than a small list of companies that could each become a single point of failure.

Her portfolio didn't become "more aggressive." It became more *intentional*. That reduced anxiety more than her oversized cash pile ever did, because her money finally had a purpose she could explain.

The most surprising part for her was that she felt calmer even though she held more stocks than before. The calm didn't come from lower volatility; it came from better design. She knew her cash buffer could handle a layoff, so she no longer needed her retirement account to behave like an emergency fund.

The most common mistake: using the wrong tool for the job

Late starters often make one of two errors:

- They use *cash* for long-term growth because it feels safe.
- They use *stocks* for short-term needs because they're chasing return.

Both create preventable stress.

If you need money in the next year or two—for a home repair, a car replacement, a planned tuition bill—that money shouldn't be exposed to stock market mood swings. Even a great stock fund can be down at the wrong moment, and your timing need doesn't care about the market's long-term average return.

On the other hand, if you're trying to fund decades of retirement spending, you can't ask cash to do the work of stocks. Cash is not an engine. It's storage.

Bonds sit in the middle. They can support both sides: they're generally more stable than stocks, and they usually offer better long-term potential than cash. But they are not immune to inflation or interest-rate changes. Bonds are stabilizers, not a guarantee.

A quick decoder ring for confusing fund labels

Fund names are often marketing. What matters is what the fund *holds*.

When you see:

- *"Growth" or "Equity"*: usually stock-heavy.
- *"Income"*: usually bond-heavy (or a mix), aiming for interest and dividends.
- *"Balanced"*: usually a mix of stocks and bonds.
- *"Money Market"*: cash-like.
- *"Alternative," "Managed Futures," "Option Income," "Structured"*: these can be complex; read carefully before treating them as "safer."

If you only remember one rule: before you buy anything, ask, "Is this gas, brakes, or airbag?" If the answer is unclear, keep looking.

Actionable takeaway: give every dollar a job

Pick one account statement—your 401(k), IRA, or brokerage—and label each holding as *Stocks, Bonds*, or *Cash*. If you can't label it quickly, that's your cue to read what it owns (or simplify).

Then write a one-sentence job description for each bucket:

- Cash: "This keeps me from selling investments during a crisis."
- Bonds: "This reduces the portfolio's stomach-dropping moments and supports steadier progress."
- Stocks: "This is how my money keeps up with the future cost of my life."

When every dollar has a job, market volatility stops feeling like a personal insult and starts feeling like weather. You still prefer sunny days. You just don't rebuild your house every time it rains.

4.4 The Power of Diversification

There's a particular kind of hope that shows up when retirement feels close: the hope that one great pick will fix everything.

It's understandable. If you feel behind, the slow-and-steady approach can sound like a lecture from someone who started early. Meanwhile, your brain is doing the math it *wants* to do: "If I could just find the next Apple, the next Nvidia, the next whatever, I could catch up without making painful changes."

That's the story of the needle.

Diversification is the boring-sounding alternative: buy the haystack. Own a little bit of everything, accept that you won't hit the jackpot, and trust the global economy to keep doing what it has always done—grow, innovate, stumble, recover, and grow again.

For late starters, diversification isn't a conservative choice. It's an aggressive form of self-protection. When time is short, you can't afford the kind of mistake that takes a decade to heal.

The true goal: avoid the unrecoverable

Investing after 40 changes what matters most. You still want growth, maybe more than ever. But you don't just need growth; you need *survivable* growth.

A 25-year-old who makes a catastrophic mistake has time to recover. A 55-year-old who makes a catastrophic mistake has a new life plan.

That's why the most dangerous risk is not "my portfolio will have a bad year." Bad years are expected. The most dangerous risk is "one decision will permanently shrink my future."

Concentration—putting too much into one company, one sector, or one country—is how that permanent shrink happens. It's the financial equivalent of building your house on one stilts' worth of ground.

Diversification is how you spread the load-bearing weight across many supports.

What diversification really is (and what it isn't)

Diversification doesn't mean owning a lot of stuff. It means owning stuff that behaves differently.

You can own ten funds and still be undiversified if they all move together. Many portfolios that look "busy" are secretly one bet: U.S. large-cap growth stocks wearing different outfits.

Real diversification asks a more interesting question: *What happens if my favorite thing disappoints?*

Because disappointment is not rare. It's normal. Companies miss earnings. Industries get disrupted. Regulations change. Consumer tastes shift. The world rarely announces when it is about to change its mind.

Diversification is the deliberate refusal to need any single prediction to be right.

The investing world is a graveyard of "sure things"

If you've been around long enough to remember a few market cycles, you've seen the pattern:

At any given moment, there is a dominant narrative—a sector that feels inevitable, a group of companies that seem unstoppable, a theme that becomes shorthand for "the future." Then the future arrives and it looks different than expected. Sometimes the theme still wins, but the winners change. Sometimes the theme was correct but overpriced. Sometimes it was just a story.

The graveyard is full of companies that were once considered permanent:

- iconic retailers that couldn't adapt to online shopping,
- household-name tech firms that got leapfrogged,
- "can't miss" energy and commodity booms that crashed when demand shifted,

- banks that looked like fortress institutions until they didn't.

The point isn't cynicism. It's humility. The market is a machine that turns certainty into surprises.

When you diversify, you're not claiming to know what will win. You're admitting you don't—and building a portfolio that doesn't require you to know.

A simple metaphor: you're buying a harvest, not a seed

Stock picking is like choosing a seed and saying, "This one plant will feed me."

Diversification is like buying a harvest from many farms. Some will have droughts. Some will have record yields. Some will switch crops. You're not betting your dinner on one field.

This matters even more when time is short. A concentrated portfolio can grow quickly, yes. It can also crater quickly. And a crater at 58 doesn't just change numbers; it changes choices: when you retire, where you live, whether you can help family, whether you can say no to work that's hurting you.

Diversification is how you protect your choices.

Case study: the employee-stock trap

A classic concentration problem is the employee who is unintentionally betting their entire financial life on one company.

Take "Paul," 54, who works at a publicly traded company that has treated him well. He has:

- a paycheck from the company,
- health insurance from the company,

- a big chunk of his 401(k) invested in the company stock (because it has done well),
- and some stock options or restricted stock that will vest over time.

Paul doesn't think of this as concentration. He thinks of it as loyalty and familiarity. The company feels like something he understands. It feels less risky than "the market."

But look at the risk profile: if the company hits trouble, Paul can lose his job *and* watch his retirement savings fall at the same time. That's not diversification. That's stacking the same dominoes.

This isn't hypothetical. Corporate histories are full of moments where a company's stock dropped 50%, 70%, even more—and layoffs followed. Sometimes it was fraud. Sometimes it was competitive pressure. Sometimes it was a CEO who made one bad acquisition with too much debt. The details vary; the pattern doesn't.

For Paul, the fix wasn't dramatic. It was systematic:

- He set a maximum percentage of total net worth he was willing to keep in employer stock.
- He made a plan to sell shares as they vested and redirect into a broad index fund.
- He separated "belief in the company" from "betting his retirement on the company."

That last step is the emotional work. Many people feel disloyal reducing employer stock exposure. But your retirement is not a tribute. It's your future paycheck.

Why "picking winners" is harder than it sounds

Most people don't realize they're signing up for two jobs when they pick stocks.

Job one is finding the right company.

Job two is finding the right price.

You can be right about the company and still lose money if you overpay. A great business can be a terrible investment at the wrong valuation. That's not a beginner mistake; it's a permanent feature of markets.

There's also the reversion problem: when something has been winning for a while, it becomes more expensive, more popular, more celebrated. It feels *safe* because it's familiar and because everyone else agrees.

This is how concentrated bets get built: not out of recklessness, but out of a very human tendency to confuse recent success with permanent success.

Diversification is a way of refusing to buy only what has been winning lately.

Diversification and the miracle of "not needing to be right"

A diversified portfolio doesn't require you to answer questions like:

- Will U.S. stocks outperform international stocks in the next decade?
- Will technology keep dominating?
- Will interest rates stay high?

- Which companies will survive the next wave of disruption?

Instead, it makes a simpler bet: that the global economy will produce winners, even if you can't name them in advance.

That bet has a track record. Not a guarantee, not a law of physics, but a historically consistent pattern: human beings solve problems, create value, and get paid for it. Markets reflect that messy progress over time.

Diversification is how you tap into that pattern without needing to win the prediction contest.

The "haystack" approach: what it looks like in practice

Buying the haystack often means owning broad funds that cover:

- a wide slice of U.S. stocks,
- a wide slice of international stocks,
- and high-quality bonds for stability.

Those three buckets can hold thousands of securities across industries and countries. You get exposure to innovation without needing to guess the innovator. You get exposure to global growth without needing to pick the right continent.

This approach can feel unsatisfying to people who want a story. It's not a cocktail-party strategy. It doesn't give you a thrilling opinion.

But it gives you something better: a plan that keeps working even if your opinions are wrong.

The downside of diversification (and why it's still worth it)

Diversification has one emotional cost: it guarantees you will always own something disappointing.

When U.S. stocks surge, your international holdings might lag. When stocks soar, your bonds might look boring. When the market is obsessed with one sector, your broad index fund will look like it's missing out.

That discomfort is the price of not needing a hero.

And it's a good price to pay, because the alternative is far worse: owning only the recent winners, then discovering that "recent" is not the same as "forever."

A diversified portfolio is not designed to maximize bragging rights. It's designed to minimize the chance you have to start over at 57.

A quick test: are you diversified or just crowded?

You don't need a finance degree to diagnose concentration. Ask a few blunt questions:

- If one company went to zero, would it ruin my retirement?
- If one sector crashed (tech, energy, real estate), would I be seriously damaged?
- If the U.S. market had a long flat decade, do I have exposure elsewhere?
- If inflation stayed high, do I have a plan beyond hoping it stops?

If any single answer makes you wince, diversification is not a philosophical debate. It's a practical repair.

The late-starter advantage: you can diversify faster than you think

Late starters sometimes assume they can't diversify because their portfolio is "too small" or their choices are limited. In reality, diversification is often easier than stock picking because it can be done with a handful of broad funds.

You don't need twenty positions to be diversified. You need coverage: many companies, many industries, multiple countries, and a mix of growth and stability.

You can also diversify with new money. If you have a concentrated position you don't want to sell immediately (because of taxes or emotions), you can direct future contributions toward other areas until the concentration naturally shrinks as a percentage of the whole.

Diversification is not an all-or-nothing switch. It's a direction.

Actionable takeaway: set a concentration limit and honor it

Pick one number: the maximum percentage of your total investments you're willing to have in any single stock or any single sector. Many people choose something like 5% for a single stock and a modest cap for any sector, but the exact number matters less than the act of setting a boundary.

Then do one concrete thing this week:

- If you're already over your limit, redirect new contributions toward broad funds until you're back inside it.
- If the concentration is in employer stock, set an automatic plan to sell vested shares and diversify.

- If your portfolio is "crowded" but not diversified, replace overlapping funds with a simpler, broader holding.

The goal isn't to eliminate risk. The goal is to eliminate the kind of risk that can't be repaired on a deadline. When you buy the haystack, you stop needing a needle to save you. You just need time, consistency, and a portfolio built to survive reality.

4.5 Low-Cost Indexing Strategies

If you want to see how the financial industry thinks about you, look at the part of the statement most people ignore: fees.

Not the dramatic kind—not the "I paid $1,200 for a planner meeting" kind. The quiet kind. The kind that slides out of your account in fractions of a percent, year after year, with all the emotional impact of dust.

Fees are where the game gets unfair in a very specific way: they are one of the only forces in investing that are nearly guaranteed. Markets might boom, crash, or churn sideways. Your fund's expense ratio will show up like rent.

And when you're investing after 40, those guaranteed drips matter more. You have less time for recovery, less time for compounding to overcome mistakes, and less patience for strategies that require a lucky decade to justify their cost.

Keeping costs low isn't a *nice-to-have*. It's a catch-up tool.

The fee you don't feel is the fee that does the most damage

A fund's *expense ratio* is the annual fee charged by the fund itself, expressed as a percentage of the assets you have invested. If a fund has an expense ratio of 1.5%, you won't typically get an invoice for 1.5%. You'll simply earn 1.5% less than you otherwise would have, because the fund removes its costs from the fund's assets.

That subtlety is what makes fees dangerous. Humans respond to pain we can see. A $20 late fee on a credit card is infuriating. A 1.5% drag on your retirement portfolio is invisible—and far more expensive.

It's also psychologically tricky because 1.5% sounds tiny. Most people can't feel the difference between 6% and 7.5% in a single year. Over 20 years, it can be the difference between "retire with dignity" and "work longer than you planned."

The market doesn't send you a sympathy note when you overpay. It just compounds whatever you leave on the table.

Why fees are a double tax on your future

A fee doesn't just reduce your return once. It reduces:

- the money you have at the end of the year, and
- the base that gets to compound next year.

So the real cost of fees is not the fee you paid. It's the growth you *never got to earn* because that money was siphoned off before it could multiply.

That's why a 1% difference in annual fees can create a shockingly large gap. It's not drama. It's math behaving consistently.

The wall-street secret hiding in plain sight

Here is a statement that sounds cynical until you think about it for ten seconds:

> The average investor in "active" funds can't beat the market after fees, because together they *are* the market.

Active management means a professional manager is selecting investments and trying to outperform a benchmark index (like the S&P 500). Passive investing (indexing) means you're simply buying the benchmark itself, trying to match it at low cost.

Before costs, the collection of all active investors, taken together, must earn roughly the market return—because they collectively own the market's stocks and bonds in some proportion. After costs (fees, trading costs, taxes created by turnover), the average active dollar has to lag.

That doesn't mean *no* manager can outperform. Some will. The problem is that you can't know in advance which ones, and the ones that did well in the past don't reliably repeat in the future. Meanwhile, the fees are absolutely reliable.

Indexing is not an admission that you don't care. It's a decision to stop paying extra for a prediction contest you don't need to enter.

A concrete example: the same portfolio, two different fee levels

To make this real, imagine two people with identical behavior:

- Both invest $10,000 today.
- Both add $6,000 per year for 20 years.
- Both earn the same *gross* market return before fees (assume 7% annually for illustration).

The only difference is cost:

- Investor A uses low-cost index funds with 0.1% annual fees.
- Investor B uses higher-cost actively managed funds with 1.5% annual fees.

That 1.4% fee gap doesn't just reduce returns; it reduces the compounding engine itself. The difference can easily become tens of thousands of dollars. In many reasonable scenarios, it becomes well into six figures.

And here's the gut-punch: to overcome a 1.4% fee disadvantage, the expensive portfolio has to outperform *every year* just to tie. Not once. Not "on average." Consistently enough to offset the drag. That's a high bar in a world where even the pros are competing against each other.

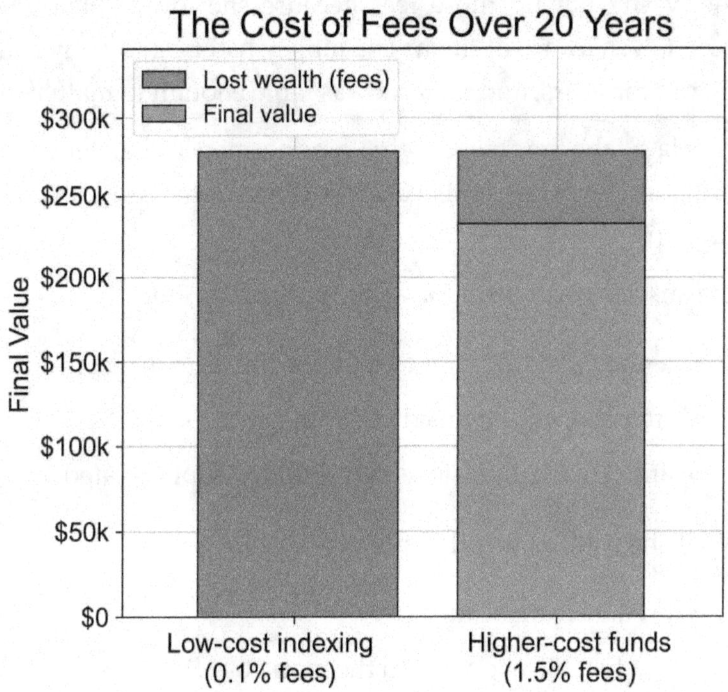

Assumes $10,000 initial + $6,000/yr, 7% gross return, 20 years

The case study nobody brags about: Angela's "boring" turnaround

Angela was 47 when she finally looked closely at the funds in her old 401(k). Like a lot of people, she had picked something years ago based on a short description and a star rating. Her portfolio wasn't outrageous, but it was expensive: multiple actively managed funds with expense ratios around 1% to 1.8%.

She wasn't paying those fees because she loved them. She wasn't even aware of them. The money had been quietly leaking out for years, and the leak was large enough to matter.

Angela's "aha" moment wasn't a performance chart. It was a question: "If these funds cost this much, what am I actually buying?"

The answer was uncomfortable. She was buying:

- a manager's attempt to outguess other managers,
- frequent trading inside the fund,
- and a marketing story that sounded sophisticated.

What she wanted was different. She wanted:

- broad diversification,
- predictable exposure to the market,
- and the highest possible chance of keeping the return the market *already offers*.

So she switched to a low-cost index lineup: a broad U.S. stock index fund, an international stock index fund, and a bond index fund. Nothing fancy. Nothing to talk about at parties.

Her investing experience changed immediately. Not because the market changed, but because her relationship to the plan changed. She stopped wondering whether her manager was "still good." She stopped chasing the latest "top fund" list. She started thinking in decades again.

The most interesting part: Angela reported feeling *less* anxious, even though her portfolio still went up and down. The anxiety had been fueled by a sense that she needed to keep making smart calls. Indexing removed the burden of constantly needing to be clever.

What "indexing" actually means in a retirement account

Indexing is not a special product. It's a design choice:

- Own the market rather than trying to beat it.
- Keep costs low.
- Keep trading low.
- Stay consistent.

In practice, that usually means using mutual funds or ETFs that track broad indexes. These funds can cover thousands of companies with a single purchase. They don't have to pay a manager to research and trade constantly, so they can charge less.

In a 401(k), you'll often see index options labeled something like:

- "S&P 500 Index" (large U.S. companies),
- "Total Stock Market Index" (broad U.S. stocks),
- "Total International Stock Index" (non-U.S. stocks),
- "U.S. Bond Index" (broad investment-grade bonds).

In an IRA or taxable brokerage account, you'll see similar choices, sometimes as ETFs.

The names aren't important. The structure is: broad exposure, low cost, no drama.

The three-fund idea: simple enough to live with

A common low-cost indexing approach is the so-called "three-fund portfolio," which uses three broad funds:

- U.S. stocks,
- international stocks,
- bonds.

The genius is not in the number three. The genius is that it's hard to mess up.

A portfolio that is simple is more likely to be held through ugly years. A portfolio that is complicated tends to invite fiddling, and fiddling is how long-term plans get sabotaged. Complexity makes people feel productive. Indexing makes people wealthy quietly.

You can absolutely build other variations—one global stock fund instead of two, a different bond mix, a target-date fund that bundles everything. The point is to stop confusing "more moving parts" with "better."

Common fee traps that feel reasonable in the moment

Some higher-cost options don't look like "fees" at first glance:

- *Actively managed funds with persuasive narratives*: the manager is "defensive," "tactical," "risk-managed," or "opportunistic." Sometimes those words hide a high expense ratio and heavy trading.

- *"No-load" funds*: "No-load" only means no upfront sales commission. The fund can still have a high expense ratio.
- *Variable annuities and complex wrapped products*: these can layer multiple fees (mortality and expense charges, admin fees, underlying fund fees, rider fees). They may have valid uses in specific situations, but they are not automatically "safer." They are often *expensive*.
- *Frequent trading inside taxable accounts*: even if your fund fee is low, turnover can create taxable distributions that reduce your net return.

The pattern is consistent: when a product is sold on a story, check the price tag.

How to shop for low-cost funds without becoming an expert

You don't need to read a 120-page prospectus for fun. You just need to know where to look.

In a workplace plan, find the fund list and look for:

- expense ratio (sometimes abbreviated as "ER"),
- whether it's an index fund,
- the benchmark it tracks.

In an IRA or brokerage, the same fields apply.

As a rough gut-check, broad index funds are often available at very low expense ratios (sometimes well under 0.20%). Many actively managed funds charge 0.75% to 1.50% or more. The exact cutoffs aren't the point. The point is to notice when you're paying premium prices for ordinary exposure.

Also remember: a fund with higher fees isn't "bad." It's just *required to be better* to justify itself. For most investors, the odds are better when you choose the option that doesn't need heroics.

The emotional benefit of indexing: fewer decision points

Indexing doesn't just reduce fees. It reduces the number of times you're tempted to make a costly decision.

Active strategies invite second-guessing:

- "The manager is underperforming—should I switch?"
- "This sector fund is hot—should I add it?"
- "This new strategy seems safer—should I rotate?"

Every one of those questions creates an opportunity to buy high and sell low. Indexing is not passive in the sense of "doing nothing." It's passive in the sense of "refusing to turn investing into a weekly referendum on your intelligence."

For late starters, reducing decision fatigue is a legitimate financial advantage. The best plan is the one you can follow when you're tired, busy, and slightly scared.

Actionable takeaway: run a fee audit and cut one leak this week

Pick your largest account and do a simple audit:

- List each fund and its expense ratio.

- Circle anything above 0.50% and ask, "What am I getting for this?"
- If the answer is vague—or if the fund is simply "the market, but pricier"—replace it with a low-cost index alternative if available.

Then make one change that sticks: set your default contributions to the low-cost lineup (or a low-cost target-date fund if you want the all-in-one option). The market will still do what it does. The difference is that you'll keep more of whatever it gives you, and your future self will not have to fund someone else's yacht to pay for your own retirement.

4.6 Designing Your Asset Allocation

Asset allocation is the part of investing that feels suspiciously like adulting: you make a choice now that your future self will either thank you for or glare at you about.

It's also where late starters tend to swing between two bad extremes. Some people go all-in on stocks because they feel behind and want to "make it up." Others go all-in on safety because the idea of a big market drop feels unbearable. Both approaches are emotional. Neither is a plan.

A workable asset allocation after 40 is a compromise between two truths that don't like each other:

- You need growth to outrun inflation and build a bigger future.
- You need stability so you don't abandon the plan during the inevitable ugly years.

The right mix is not a moral category. It's a design decision.

Start with the only question that matters: what can break you?

A portfolio doesn't fail because it has a bad year. It fails because you do something irreversible in response to a bad year.

So before picking percentages, name the conditions that could force you to sell when prices are down:

- a job loss,
- a medical bill,
- a family emergency,
- a sudden need to move,
- a large debt payment that can't be delayed.

If you don't have a buffer for those events, you're not "conservative." You're fragile. Fragility is what turns a temporary market decline into a permanent life change.

Asset allocation is easier when your financial life has shock absorbers: an emergency fund, manageable debt, and insurance that actually covers what it's supposed to cover. With those in place, you can invest for the timeline you *have*, not the timeline your anxiety insists on.

Time horizon is not your age; it's your first withdrawal

A 58-year-old who plans to work until 70 has a longer time horizon than a 50-year-old who wants to retire at 57. Age is a rough clue. Your real horizon is the moment you'll need to start taking money out.

That first withdrawal matters because it changes the game. When you're only contributing, market drops are uncomfortable but not fatal. When you're withdrawing, drops can become dangerous if you're forced to sell low to fund living expenses. That's why the closer you are to spending from the portfolio, the more important it becomes to have some stability built in.

But "stability built in" doesn't mean "give up on growth." It means being intentional about the role of each asset class:

- Stocks: long-term growth, inflation-fighting power.
- Bonds: shock absorption, income, a counterweight to stock volatility.
- Cash: immediate needs, panic prevention, flexibility.

The allocation question becomes: how much growth can you hold through stress without flinching, and how much stability do you need so you can sleep?

Risk tolerance isn't bravery; it's behavior

The most common investing lie people tell themselves is: "I'm fine with risk."

What they usually mean is: "I'm fine with risk in my imagination."

Risk tolerance isn't what you say when markets are calm. It's what you do when your portfolio is down 20% and everyone you know is suddenly an economist. It's your ability to continue contributing, keep your job separate from your investments, and not punish your future because the present feels scary.

If you don't know your risk tolerance, look at your history:

- Did you sell during a downturn?
- Did you stop contributing?
- Did you change funds repeatedly based on recent performance?

Those actions aren't shameful. They're information. They tell you what kind of portfolio you can actually hold.

The "best" asset allocation on paper is useless if it causes you to bail out at the bottom. A slightly more conservative allocation that you can stick with often wins in real life.

The policy matters more than the picks

There's a reason serious institutions spend so much time debating asset allocation. Over long periods, the broad mix of stocks and bonds tends to drive the bulk of results. The fund-of-the-month matters far less than whether you chose an allocation you can live with.

That's good news for late starters. It means you don't need to find a genius manager or the perfect stock. You need a policy—a default mix—and the discipline to maintain it. Investing becomes less like gambling and more like engineering.

Three templates that work in the real world

People often want one "correct" allocation. What actually works is picking a reasonable template and committing to it long enough for the math to do its job.

Here are three starting points that many late starters use. They're not commandments. They're scaffolding.

Aggressive catch-up (example: 80/20 or 75/25). Mostly stocks, with a smaller bond allocation. This is for someone who:

- has a solid emergency fund,
- has stable income,
- won't need to tap the portfolio for many years,
- can tolerate large drops without changing course.

The upside is higher expected growth. The downside is bigger and more frequent stomach-churning declines.

Moderate growth (example: 70/30 or 60/40). A classic balance: enough stocks to grow, enough bonds to steady the ride. This is for someone who:

- wants growth but knows they will panic if the portfolio feels like a roller coaster,
- is within 10–20 years of retirement,
- values staying invested more than swinging for the fences.

For many investors over 40, this becomes the "sweet spot" because it's hard to abandon.

Conservative transition (example: 50/50 or 40/60). More bonds than stocks. This is for someone who:

- is close to retirement,
- expects to begin withdrawals soon,
- has lower tolerance for volatility,
- prioritizes reducing the impact of a major market decline.

The tradeoff is that this allocation may struggle more against inflation over long retirements, so the withdrawal plan and spending flexibility matter a lot.

These templates are easiest to implement using low-cost index funds: a broad U.S. stock fund, a broad international stock fund, and a broad bond fund. The precise split between U.S. and international stocks can vary; the main lever is stocks versus bonds.

Case study: Kevin and the portfolio that stopped him from quitting

Kevin was 44, recently divorced, and had the familiar late-starter cocktail: a smaller-than-he-wanted retirement account and a larger-than-he-wanted sense of urgency. He did what urgency often pushes people to do: he built a stock-heavy portfolio because it seemed like the only way to catch up.

Then 2022 happened, and his account dropped hard enough that he started doing late-night math. Not the empowering kind. The grim kind. He didn't sell, but he stopped contributing for six months because it felt pointless to "throw money into a falling market." That pause cost him more than it seemed: not just missed contributions, but missed opportunities to buy when prices were lower.

What fixed it wasn't a better forecast. It was a portfolio he could hold.

Kevin switched to a moderate allocation: 70% stocks, 30% bonds. The bonds didn't guarantee profits. They gave his portfolio a less violent personality. When stocks dropped, the whole account didn't feel like it was in free fall. Kevin started contributing again, because the plan no longer required him to be emotionally heroic.

Over time, the change in behavior mattered more than the change in percentages. A plan you fund beats a plan you abandon.

Bonds: stability isn't one size fits all

Once you decide you want bonds, you still have choices about what kind of stability you want.

High-quality bond funds (often heavy in government and investment-grade bonds) tend to be used for ballast. They can still drop when rates rise, but they usually behave more predictably than lower-quality credit.

High-yield bonds often behave more like stocks when stress hits, because credit risk becomes the story. They can have a place, but they are not the same "brakes" as high-quality bonds. If your bond allocation is meant to be the calm part of the portfolio, keep it truly calm.

Duration matters too. Longer-duration bonds can swing more when interest rates move. Shorter-duration bonds tend to be less sensitive. If your goal is stability, you may prefer a bond mix that doesn't amplify rate risk.

The goal with bonds in a catch-up plan is not to squeeze out maximum yield. It's to prevent your stock allocation from being sabotaged by panic.

Cash: the most underrated part of asset allocation

Most allocation discussions pretend cash doesn't exist, as if adults never have emergencies.

Cash is what keeps you from using your retirement account as a crisis ATM. It's also what keeps you from selling at the wrong time.

If you're within a few years of retirement, some people also build a small "spending runway" in cash or short-term, high-quality bonds—money that can cover near-term spending so you aren't forced to sell stocks after a market drop. This isn't about timing the market. It's about buying yourself choices.

The exact amount depends on your job stability, your household needs, and how much volatility you can tolerate. The principle is simple: use cash to protect your long-term assets from short-term demands.

How to choose your allocation in 30 minutes (without a personality quiz)

If you want a practical way to choose an allocation without getting lost in theory, try this:

Step 1: Pick your "worst year" number. Ask: "If my portfolio dropped by __________%, would I stay invested and keep contributing?" Common stress tests are 20%, 30%, even 40%. If 30% makes you feel physically ill, don't build a plan that requires you to endure 30% without blinking.

Step 2: Translate that into a stock percentage. More stocks generally means bigger drops in bad years. More bonds generally means smaller drops but lower expected long-term growth. You're not trying to engineer the perfect outcome. You're choosing the mix that keeps you in the game.

Step 3: Commit to a range, not a point. Instead of "I am 65/35 forever," use a band: "I'm roughly 60–70% stocks." That gives you flexibility while keeping your plan anchored.

Step 4: Automate and stop negotiating with yourself. Once your allocation is set, set contributions to flow into the funds that maintain it. The more you can make the plan boring, the more likely you'll keep it.

When simplicity wins: the target-date fund option

If you want a professionally managed allocation that automatically becomes more conservative over time, a low-cost target-date fund can be a strong choice inside a retirement account. You pick a fund with a date near when you plan to retire. The fund holds a diversified mix of stocks and bonds and gradually shifts the mix as the target date approaches.

Target-date funds vary widely in cost and design, so the same fee discipline still applies. But for many late starters, the real value is behavioral: it removes the temptation to constantly redesign your portfolio based on fear or headlines.

Actionable takeaway: choose an allocation you can fund through a downturn

Pick one of the templates (aggressive catch-up, moderate growth, conservative transition) and set it as your default for the next 12 months. Then do one stress-test exercise that's uncomfortably honest:

Write down the dollar amount your portfolio would lose in a 25% decline, based on your current balance. Sit with the number. If it makes you feel like you'd sell, reduce the stock allocation now—while you're calm—instead of later when you're scared.

The "right" allocation is the one that lets you keep contributing when the market is loud. Consistency is the real edge, and the only portfolio that compounds is the one you can hold.

4.7 Staying on Track: Rebalancing

A portfolio doesn't drift because you did something wrong. It drifts because it worked.

That sounds like a motivational poster, but it's also literal math. If stocks have a great run, the stock portion of your portfolio becomes a bigger slice of the pie. If bonds have a stretch of strength, the bond slice swells. Your original mix—the one you chose because it matched your timeline and your nerves—slowly morphs into something else.

Left alone, a portfolio quietly rewrites its own risk level.

And after 40, "quietly rewriting risk" is not a cute feature. It's how people accidentally end up with a portfolio that's too aggressive right before a market drop, or too conservative right before the years they desperately need growth.

Rebalancing is the maintenance that keeps the plan from turning into a stranger.

What rebalancing actually is (no mysticism required)

Rebalancing means bringing your portfolio back to your target allocation—your chosen mix of stocks, bonds, and cash—after market movements knock it off course.

If your target is 60% stocks and 40% bonds and a strong stock market pushes you to 72/28, rebalancing means moving money from stocks to bonds until you're back near 60/40.

The mechanical action can be:

- selling some of what grew faster and buying what lagged,
- directing new contributions into the underweight asset (often the simplest),
- or, in some accounts, exchanging between funds.

The emotional action is even more important: rebalancing forces you to do the opposite of your instincts. When something has been soaring, it feels "safe" and exciting, and you want more of it. Rebalancing says: take some off the table. When something has been disappointing, it feels "broken," and you want to abandon it. Rebalancing says: add to it.

It's the closest thing investing has to a built-in anti-panic mechanism.

Why drift is dangerous when time is short

Portfolio drift is easiest to ignore during good years, which is exactly when it becomes most dangerous.

Imagine you're 52, targeting retirement in your mid-to-late 60s. You set a moderate allocation because you want growth but don't want your stomach to run your finances. A few strong equity years go by and your portfolio becomes stock-heavy.

If the market then drops hard, you don't experience the drop you *planned* for. You experience the drop you *drifted* into.

That difference matters because your tolerance for volatility isn't about percentages in a spreadsheet. It's about the lived experience of watching a number you depend on get cut down. Drift can push you over your emotional limit, and the moment you cross that limit is the moment you do something irreversible.

Rebalancing is less about squeezing out extra return and more about keeping your portfolio within the range you can hold.

Case study: Sharon's "accidental 90% stock portfolio"

Sharon, 56, was the kind of investor advisors love: steady contributions, no silly fads, no day trading. Years earlier, she had chosen a 70/30 portfolio in her IRA, aiming for a balance of growth and stability.

Then the market had a long, strong run. Sharon didn't touch anything because she assumed "not touching" was disciplined. Her quarterly statements looked great. She felt smart without doing anything.

When she finally looked under the hood—prompted by a friend's casual question, "What are you allocated to these days?"—she discovered she was close to 90% in stocks. Her bond fund hadn't disappeared; it had simply been outgrown.

Sharon's risk hadn't increased because she became reckless. It increased because her winners won.

She decided to rebalance in two steps:

- She redirected all new contributions to bonds until her mix moved closer to target.
- She sold a portion of stocks inside the IRA (no tax impact for trades inside the IRA) and bought bonds to get back to 70/30 quickly.

A few months later, the market dipped. Sharon still didn't enjoy it, but she didn't panic. Her portfolio felt like the one she had agreed to live with. That's the difference rebalancing makes: it turns market drama into something you can endure.

The rebalancing paradox: it feels like you're doing the wrong thing

Rebalancing asks you to sell what has been doing well and buy what has been doing poorly. That can feel like sabotaging yourself.

In a world where everyone cheers winners, rebalancing feels like a refusal to celebrate.

But the point isn't celebration. It's control.

A portfolio is not a scoreboard. It's a machine designed to fund your life. Machines need calibration. If you never recalibrate, the machine starts doing something other than what you intended.

Rebalancing is calibration.

It also quietly enforces a powerful behavior: "buy low, sell high." Not in a market-timing way, but in a rules-based way. You're not predicting that the lagging asset will surge tomorrow. You're simply ensuring you keep owning enough of it to benefit when it eventually has its season.

Two sane ways to rebalance (and why sanity matters)

There are many rebalancing strategies, but most people don't need sophistication. They need a rule they can follow without turning investing into a second job.

Two common approaches work well:

Calendar rebalancing.

Pick a schedule—every six months, once a year, perhaps on your birthday or at tax time—and rebalance then. The benefit is simplicity. The downside is that drift can get large between rebalances in fast markets.

Threshold rebalancing.

Rebalance when your allocation moves beyond a preset band. For example, if your target is 60% stocks, you might rebalance if stocks rise above 65% or fall below 55%. The benefit is that it responds to meaningful changes and avoids needless tinkering. The downside is you have to check occasionally, and you have to decide your bands.

Both methods are valid. The real enemy is not choosing "the perfect" trigger. The enemy is doing nothing until the market forces your hand.

If you love simplicity, pick annual. If you like a little more precision, pick thresholds. Either is better than improvisation.

How to rebalance without making taxes your enemy

Rebalancing in a tax-advantaged retirement account (like a 401(k) or IRA) is usually straightforward because buying and selling inside the account generally doesn't create current tax bills. That makes these accounts great places to do the "heavy lifting" of rebalancing.

In a taxable brokerage account, selling appreciated investments can trigger capital gains taxes. That doesn't mean "never rebalance," but it means you want to rebalance thoughtfully.

Practical ways to rebalance in taxable accounts:

- *Use new contributions*: buy more of the underweight asset instead of selling the overweight one.
- *Use dividends and interest*: direct distributions into whichever asset is underweight.
- *Harvest losses when appropriate*: in some situations, you can sell at a loss and offset other gains (this can be helpful but has rules).

The headline is simple: if you can rebalance inside retirement accounts, do it there. Keep taxable accounts as simple and tax-efficient as possible unless you have a reason not to.

Also consider transaction costs, especially in plans that still charge for trades. Many modern brokerages offer commission-free trading, but some retirement plans have restrictions or fees for frequent exchanges. Rebalancing should be disciplined, not constant.

The most overlooked rebalancing tool: your paycheck

People imagine rebalancing as selling and buying. Often, the easiest form of rebalancing is just directing where new money goes.

If your portfolio is stock-heavy because stocks had a great year, you can send new contributions toward bonds for a while. That nudges the portfolio back toward target without selling anything.

This approach has two advantages:

- It minimizes taxes and transaction friction.

- It avoids the emotional sting of "selling winners."

It's also a powerful behavior trick for late starters: steady contributions become a steering wheel, not just fuel.

When rebalancing becomes a trap

There is such a thing as over-rebalancing.

If you tinker weekly, you're not maintaining a plan; you're reacting to noise. Constant rebalancing can:

- increase transaction costs (in some accounts),
- increase taxes (in taxable accounts),
- and keep your attention glued to short-term moves.

A good rebalancing rule should protect you from drift *and* protect you from yourself. It should reduce the number of decisions you make, not increase them.

If you find yourself turning rebalancing into entertainment, the cure is a slower schedule and a written policy.

The "easy button": target-date funds

If you read the word "rebalance" and immediately feel a low-grade headache, target-date funds exist for you.

A target-date fund bundles stocks and bonds into a single fund and automatically adjusts the mix over time. You pick a target year—often near when you expect to retire—and the fund follows a "glide path," typically holding more stocks when retirement is far away and gradually increasing bonds as the target date approaches.

The main benefits are behavioral:

- You don't have to manage multiple funds.
- Rebalancing happens automatically.
- The portfolio becomes more conservative over time without you needing to remember.

For many late starters, this is not laziness. It's strategic outsourcing. If managing the portfolio yourself increases the chance you'll panic or procrastinate, a target-date fund can increase your odds of staying the course.

A few cautions that keep this option smart:

- *Check the cost.* Target-date funds can be low-cost or expensive depending on the provider.
- *Understand the glide path.* Some target-date funds stay stock-heavy even near retirement; others get conservative earlier. Neither is universally "right," but you should know what you're buying.
- *Avoid doubling up.* If you use a target-date fund, it's often best as your core holding. Adding extra stock funds on the side can undo the built-in risk management.

Target-date funds don't eliminate market risk. They eliminate the operational risk of you needing to be your own portfolio manager.

Rebalancing is a way of saying "my plan is not my mood"

On the days when the market is up, your plan will tempt you to take more risk because it feels like the world is safe again. On the days when the market is down, your plan will tempt you to take less risk because it feels like danger is permanent.

Rebalancing is the disciplined refusal to let either feeling redesign your future.

It's also a small but meaningful act of maturity: you're choosing to live by the logic you set when you were calm, not the emotion you feel when the world is loud.

Actionable takeaway: write your rebalancing rule and schedule it

Choose one rebalancing rule today and put it on your calendar:

- *Calendar option:* rebalance once per year on a date you'll remember (birthday, New Year's, tax time).
- *Threshold option:* rebalance when stocks or bonds drift more than 5 percentage points from target.

Then decide *where* you'll rebalance first:

- Use your 401(k) or IRA for the main adjustments when possible.
- In taxable accounts, try to rebalance mostly with new contributions and dividends to limit taxes.

If you want to outsource all of this, select a low-cost target-date fund aligned with your retirement timeframe and make it your default destination for contributions.

A portfolio doesn't need constant attention. It needs a simple rule that keeps it from quietly becoming the wrong portfolio at the worst possible time.

Chapter 5

Supercharging Savings with Tax Advantages

If I told you there was a guaranteed return on your investment the moment you made it, you might think it was a scam. But that is exactly what tax efficiency offers. When you are racing against the clock to close the retirement gap, you cannot afford to let taxes drag on your returns. In this chapter, we will look at the tax code not as a boring rulebook, but as a treasure map. We will uncover how to keep more of what you earn, unlock 'free money' from your employer, and use special 'catch-up' privileges designed specifically for late starters like us.

5.1 Tax Basics for Late Savers

Maria is 52, a project manager, and the kind of person who keeps a spare phone charger in every bag. She is not careless with money; she's careful. But for years, "careful" meant: pay the bills, keep a few months of cash, and if anything's left, throw it at a mortgage extra payment or a brokerage account. Then she finally looked at her retirement balance and did the mental math that makes your stomach drop: there aren't many paychecks left to do the heavy lifting.

Her first instinct was the one most late savers have: "I'll just save harder." The better move was subtler: save *smarter* by letting the tax code do some of the work. Not through gimmicks or loopholes—through the plain, boring mechanics of how income is taxed and how investment returns get shaved down by taxes year after year. Once you see those mechanics clearly, a pre-tax retirement contribution stops feeling like a sacrifice and starts feeling like a negotiated discount on your own future.

The two tax rates that run your life (whether you notice or not)

Most people have one tax number floating in their head. It's usually wrong, or at least incomplete. The U.S. income tax system has *brackets*—different slices of your income are taxed at different rates. That creates two rates that matter:

- *Marginal tax rate*: the rate on the *next* dollar you earn (and, importantly, the next dollar you *deduct*).
- *Effective tax rate*: your total tax divided by your total income—your average rate after brackets, deductions, and credits.

The marginal rate is the one that decides how powerful a pre-tax retirement contribution is. When you contribute to a traditional 401(k) or similar workplace plan, you're not magically avoiding all taxes forever. You're choosing the year in which you pay them. And the year you *don't* pay them—this year—is priced at your marginal rate.

That's why the effective rate can be a distraction. If you tell a friend, "I'm in the 22% bracket," they might reply, "No way, your effective rate is like 12%." Both can be true. But your ability to turn a retirement contribution into an immediate tax break is tied to the marginal rate—the rate applied to the top slice of your income.

Here's the simplest way to picture it: your paycheck is stacked like a layer cake. The top layer is taxed the most. A pre-tax contribution slices off the top layer first.

Why a pre-tax dollar is ``cheaper" than a regular dollar

Saving money often feels like giving something up: dinner out, a vacation, a nicer car. The surprise is that a pre-tax contribution can reduce that feeling because you're saving with dollars you never fully "got" in the first place.

Suppose you're in a 22% marginal federal bracket (ignore state tax for a moment to keep the math clean). If you put $1,000 into a regular savings account or a taxable brokerage account, you need to earn $1,000 after taxes. Roughly speaking, that could require about $1,282 of pre-tax income:

$$\$1{,}282 \times (1 - 0.22) \approx \$1{,}000.$$

But if you put $1,000 into a traditional 401(k), that $1,000 comes out before federal income tax. Your take-home pay might drop by roughly $780, not $1,000, because you just reduced taxable income by $1,000 and avoided about $220 of federal tax on that slice.

So yes, you *saved* $1,000. But your lifestyle only had to absorb about $780 of pain.

This is the core reframe for late savers: a traditional contribution is not only a retirement deposit. It's also a price negotiation with the IRS. You're buying $1,000 of future wealth at a discount set by your marginal rate.

Once you add state income taxes (if applicable), the discount can be even larger. And once you add the possibility of an employer match, the discount can start to feel like a pricing error.

Tax deferral as an interest-free loan (with a catch)

A traditional pre-tax contribution is basically tax deferral: you postpone taxes until you withdraw the money. The money that would have gone to taxes stays invested for you instead.

It's useful to think of that deferred tax as an interest-free loan from the government:

- You keep extra dollars working in your account today.
- You pay taxes later, when you take distributions.

The catch is not sinister, just real: the government expects to be repaid eventually, and repayment comes out of withdrawals as ordinary income. But the loan analogy helps you see why this can be powerful even if your tax rate ends up similar later. You're borrowing at 0% interest and investing the borrowed amount for decades.

Late savers don't have decades to waste. That makes an interest-free loan invested in markets especially valuable, because time is now your scarce resource.

A concrete case: Maria chooses between taxable saving and a pre-tax 401(k)

Maria earns $110,000 and is paid biweekly. She can afford to set aside $1,000 per month. She's deciding between:

1. putting $1,000/month into a taxable brokerage account, or
2. contributing enough pre-tax to put $1,000/month into her 401(k).

If she saves $1,000 in taxable, she needs $1,000 after taxes. If her marginal federal rate is 22% and her state rate is 5% (again, simplified), her combined marginal rate is roughly 27% (not exactly additive in every situation, but close enough for intuition). Now she needs about $1,370 of pre-tax income to net $1,000:

$$\$1{,}370 \times (1 - 0.27) \approx \$1{,}000.$$

If she instead contributes $1,000 pre-tax to her 401(k), her take-home might only fall by about $730 because she avoided roughly $270 of income tax on that slice. Same $1,000 saved. Different lifestyle cost.

Now add one more detail: Maria's plan matches 50% of contributions up to 6% of pay. If she wasn't contributing enough to get the full match, the 401(k) option doesn't just discount her savings—it also adds a second stream of money she cannot get anywhere else. Suddenly, her monthly $1,000 might trigger meaningful employer dollars.

The point isn't that taxable investing is "bad." It's that late savers rarely suffer from a lack of good options. They suffer from putting their best dollars in the wrong place.

Meet ``tax drag": the slow leak you stop noticing

If marginal tax rates explain why pre-tax saving feels cheaper today, *tax drag* explains why tax-advantaged accounts can win even when you're investing in the same funds.

Tax drag is the return you *don't* get because taxes nibble at your portfolio along the way. In a taxable account, the IRS can take a bite in years when you did nothing. You can be a patient, buy-and-hold investor and still owe taxes because the investments themselves generate taxable income:

- *Interest* from bonds and savings is generally taxed as ordinary income.
- *Dividends* may be taxed (qualified dividends often at a lower rate than ordinary income; nonqualified dividends at ordinary rates).
- *Capital gains distributions* can show up when a mutual fund sells holdings inside the fund (even if you didn't sell your shares).

In tax-advantaged accounts (traditional 401(k), traditional IRA, Roth accounts), those annual drips and drizzles don't usually create a current-year tax bill. That means more of your money stays invested, compounding.

A helpful mental image: two identical bathtubs with the same faucet (investment returns). One tub has a small drain open (tax drag). You don't notice it at first, because the tub is still filling. But over years, the waterline is meaningfully lower.

A simple illustration that doesn't require a finance degree

Imagine two investors, both 45, both investing $10,000 per year for 15 years, both earning the same *pre-tax* return of 7% annually.

- Investor A uses a tax-advantaged account and pays no annual taxes on dividends/interest/gains while the money is growing.
- Investor B invests in a taxable account and, due to taxes along the way, ends up with an *after-tax* return that's effectively 6% instead of 7%.

That one percentage point doesn't sound dramatic. It sounds like a rounding error. Over 15 years, it isn't.

The difference is compounding: returns earn returns. When taxes reduce returns each year, they don't just take a bite once—they shrink the base that future growth builds on. That's why tax drag is so sneaky: it doesn't feel like a fee, but it behaves like one.

Late savers often focus on big, visible choices ("Should I be more aggressive? Should I pick a different fund?") while ignoring the quiet structural choice ("Which account type should hold these investments?"). Yet the structural choice can be the difference between progress that feels stuck and progress that finally starts to look like something.

Why this matters more after 40: you can't out-wait taxes

If you're 25 and you make a minor inefficiency, time can paper over it. At 45, 50, 55, time is not interested in helping you. That's not pessimism; it's physics.

What you do have after 40 is usually higher income than you had in your 20s. That means two things happen at once:

- your marginal rate is often higher, so the discount on pre-tax contributions is larger, and
- your taxable investments are more likely to throw off taxable income, so tax drag can be more painful.

When people say, "I'm behind, so I need higher returns," they're often chasing the wrong lever. The highest-leverage move is frequently to stop giving away returns you already earned.

The emotional hurdle: pre-tax saving feels like locking money away

A common resistance goes like this: "I don't want to tie up my money in retirement accounts. I might need it." That fear is not irrational. Life after 40 can get complicated: college bills, aging parents, health surprises, layoffs that sting more because employers assume you're expensive.

But it helps to separate *liquidity* from *tax strategy*. You can want access to cash and still use tax-advantaged accounts as your primary engine. Many households do both:

- Keep a realistic emergency fund in cash (not heroic, not fragile).
- Use retirement accounts for the dollars intended for long-term goals.

The danger is when "I might need it" becomes an all-purpose excuse to keep everything taxable. Taxable accounts are flexible, yes. But flexibility has a price tag, and the tax code collects it in small installments.

A late saver doesn't need every dollar to be flexible. They need every dollar to be purposeful.

Marginal rates: why the ``top slice" mindset changes your decisions

People often underestimate how targeted a pre-tax contribution is. They imagine it reduces taxes across the board. It doesn't. It reduces taxes on the top slice.

That top slice might be taxed at 22%, 24%, 32%, or more depending on income and filing status. If you contribute enough to drop your taxable income into a lower bracket, the later dollars of contribution may be discounted at a lower rate than the earlier dollars. That sounds like a complication, but it's actually a feature: you don't need perfect math to benefit. You just need to know that the first dollars you contribute pre-tax are often the most powerful because they're cutting into the highest-taxed slice.

It's also why the phrase "tax bracket" is misunderstood. Being "in" a bracket does not mean all your income is taxed at that rate. It means your *last* dollars are.

Once you understand that, you stop making the classic mistake: avoiding a raise or a bonus because you think it will "push you into a higher bracket" and make you poorer. Brackets don't work that way. The higher rate applies to the additional slice, not the whole cake.

A late saver doesn't need fewer opportunities. They need to keep more of each opportunity.

Tax drag isn't only about income taxes

For many households, tax drag is also about *behavior*. A taxable account makes it easier to tinker:

- Selling winners too early (creating gains sooner than necessary),
- Swapping funds in response to headlines,
- Chasing performance and realizing avoidable taxable gains.

Tax-advantaged accounts can act like a seatbelt for long-term investing. Inside a retirement account, you can rebalance, shift allocations, or change funds without triggering the immediate tax consequences that make people hesitate or, worse, make people trade in a way that's tax-expensive.

In other words: tax-advantaged accounts don't just reduce the government's cut. They can reduce your own self-inflicted cuts.

A reality check: tax-advantaged doesn't always mean ``traditional"

Pre-tax savings are powerful, but they're not the only tax-smart move. Sometimes the best option is Roth (pay taxes now, potentially withdraw tax-free later), especially if you expect higher income in the future, or if you're early in a career, or if you value tax diversification. But for many late savers currently in their peak earning years, the traditional pre-tax contribution is often the workhorse precisely because the marginal-rate discount is sitting there, waiting.

Think of it like buying groceries with a coupon. If you have the coupon and you're already shopping, using it isn't a philosophy. It's arithmetic.

The ``invisible raise": how pre-tax contributions can feel like found money

One of the most motivating moments for a late saver is realizing they can increase retirement contributions without their lifestyle collapsing.

Here's a practical way to test it. If you currently contribute nothing (or very little) to a traditional workplace plan, increase your contribution by 1% of pay and watch your paycheck. The drop is usually smaller than people fear because taxes adjust. Do it again next pay period or next month. Gradually, you can ratchet up contributions while letting the tax system cushion the landing.

The reason it works is psychological as much as mathematical. You're not trying to summon discipline out of thin air. You're redesigning your default so the money never becomes spendable in the first place.

Late savers don't need to become different people. They need a different system.

Actionable takeaway

Pick one paycheck in the next two weeks and run a tiny experiment: increase your traditional workplace retirement contribution by 1% (or set a flat dollar increase like $25—$50 per paycheck) and compare your net pay to the prior paycheck. Write down two numbers on a sticky note or in your phone:

- the amount that actually hit your retirement account, and
- the amount your take-home pay fell.

That gap—the difference between what you saved and what it *cost* your lifestyle—is the tax advantage made visible, and it's often the spark that turns "I should save more" into "I can actually do this."

5.2 Mastering the Workplace Plan

A workplace retirement plan is one of the few places in adult life where someone offers you a deal that is both legal and wildly generous: "If you put money in, we'll put money in too." No negotiating. No small print that turns the promise into a mirage. Just a policy written into your benefits package, quietly waiting to be claimed.

And yet the match is missed every day, usually for understandable reasons. People mean to enroll "after the holidays." They leave a job and mentally close that chapter. They assume they can "catch up later." Or they get overwhelmed by a menu of funds that reads like a pharmaceutical label and decide to postpone the whole thing until they feel smarter. The result is the most expensive kind of procrastination: not the cost of a late fee, but the cost of forfeited compounding *plus* forfeited free money.

After 40, you don't need a perfect retirement plan. You need a plan that captures the biggest, easiest wins first. A well-run 401(k) or 403(b) is built for exactly that.

The match: the highest-priority investment you will ever be offered

If your employer matches your contributions, the match is the first domino. It's the closest thing most people will ever see to a guaranteed return.

Here's why the match is so powerful even if your investments have a rough year: your employer's money typically shows up regardless of what the market does that month. If you contribute $100 and they match $50, you effectively start with $150. Your personal return on day one is 50% before the market even clears its throat.

People argue about whether stocks will return 7% or 9% over the next decade. No one argues that 50% is good.

The match also changes the emotional math. Saving can feel like deprivation. A match makes it feel like leverage. You're not only sacrificing spending; you're triggering an immediate payoff that can't be replicated in a taxable account, a bank account, or by buying "the right" index fund.

The only rational objection to taking the match is cash-flow reality—not preference, not skepticism, not the fear that you picked the wrong fund. If you truly can't afford it today, that's a different problem to solve. But if you *can* afford it and simply haven't turned it on, you're leaving part of your compensation unclaimed, like an uncashed paycheck.

A real-world case: Andre's ``raise" that wasn't a raise

Andre is 47 and works for a hospital system with a 403(b). He's steady, competent, and a little tired of feeling like retirement is something he's supposed to have figured out already. His salary is $84,000. The plan matches 100% of the first 3% he contributes, and 50% of the next 2%. In plain English: contribute 5% and the employer adds 4%.

Andre was contributing 2% because that felt safe. He told himself he'd increase it when life got calmer.

When he finally looked at the numbers, the "calmer later" plan didn't make sense. At 2%, he was contributing $1,680 per year, and the employer was matching $1,680 (because the first 3% is dollar-for-dollar, and he hadn't even reached it). He increased to 5%: now he contributed $4,200, and the employer added $3,360.

That extra $2,520 in employer money didn't require Andre to work more hours, negotiate a promotion, or learn to day trade. It required one online form and a willingness to tolerate a smaller paycheck.

But the part that surprised him was how *small* the paycheck drop felt compared to the contribution increase. Because his contributions were pre-tax, his take-home didn't fall by the full $2,520. Taxes adjusted. The match, meanwhile, did not ask permission. It simply appeared.

If Andre never did anything else but consistently captured that full match for the next 15 years, he would have transformed his retirement trajectory with one decision.

The match chart: how ``free money" changes the ending

The next visual is intentionally blunt. It assumes the same investments and the same market return. The only difference is whether you collect the match.

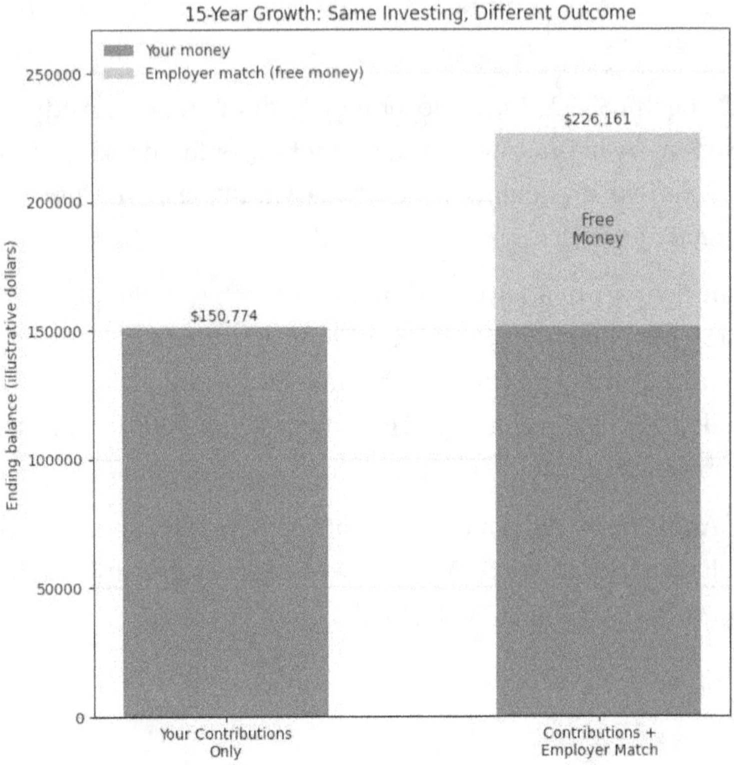

The point of a chart like this isn't to argue about whether 7% is the right return. The point is that the match changes the base your compounding sits on. Compounding is impressive; compounding on someone else's money is better.

Contribution limits: the plan has a speed limit, and it's higher than most people think

Workplace plans have annual limits on how much *you* can contribute. Those limits tend to rise over time. For 2026, the elective deferral limit for many workplace plans (401(k), 403(b), governmental 457(b), and the federal TSP) is $24,500 for most participants.

You do not need to max the plan to benefit from it. But it's useful to know the ceiling for two reasons:

- If you're trying to close a gap quickly, you need the plan's capacity. A low ceiling would be a bottleneck; a high ceiling gives you room to sprint.
- Limits force choices. If you plan to contribute aggressively, you'll want to decide early whether those dollars should be pre-tax, Roth, or a mix.

Many late savers also underestimate how quickly small percentage increases add up. On an $80,000 salary, each 1% is $800 per year. On $120,000, each 1% is $1,200 per year. Raise your contribution by 2% and you've made a move that's meaningful without being dramatic.

The practical trick is to tie contribution increases to events that already change your paycheck: annual raises, bonuses, debt payoffs, a child finishing daycare. If you wait to "feel ready," you'll be waiting for a feeling that rarely arrives on schedule.

Vesting: your money is yours; the match may have strings

One reason people treat the match as "not real" is that they've heard stories about vesting. Vesting is simply the rule that determines when employer contributions become yours to keep.

Two clarifications make vesting far less mysterious:

- *Your own contributions are always 100% yours.* If you put money in from your paycheck, you do not forfeit it.
- *Employer money may vest over time.* Some plans vest immediately. Others vest in steps (graded vesting) or all at once after a set period (cliff vesting).

After 40, vesting matters because job changes can be more consequential. If you're considering leaving an employer and you're close to a vesting milestone, the match isn't just a perk—it's a number you can calculate.

If you're 80% vested and you leave now, you may forfeit 20% of the employer portion. That may or may not change your decision, but it should be a conscious choice, not an accidental donation.

The simplest action here is also the most underused: read your plan's vesting schedule. It's not a 90-page legal document; it's usually a short table in the Summary Plan Description (SPD) or benefits portal. If you can read a restaurant menu, you can read a vesting schedule.

Traditional vs. Roth inside the plan: don't let the perfect choice block the good one

Many workplace plans offer both traditional (pre-tax) and Roth (after-tax) contributions. People get stuck here because it sounds like a high-stakes prediction: will your tax rate be higher or lower later?

A better way to think about it is this: the match is usually the first priority regardless, and the contribution type is the tuning knob.

If you're in peak earning years and your budget is tight, traditional contributions often let you contribute more with less impact on take-home pay. If you expect higher income later, value tax-free withdrawals, or want more flexibility around future tax planning, Roth contributions may be appealing.

But the bigger danger is paralysis. Contributing something (and capturing the match) beats endless deliberation. If you're uncertain, a mix can be a perfectly rational compromise: you're building tax diversification, not betting your entire retirement on a single tax forecast.

The plan menu: how to pick funds without becoming an amateur economist

Workplace plans can feel like they were designed by someone who hates you. Twenty-five funds with similar names. Expense ratios in tiny print. Performance charts that invite you to chase last year's winner.

You don't need brilliance here. You need a process that avoids the two classic mistakes: (1) doing nothing and sitting in cash for years, or (2) picking the most exciting-sounding fund and treating it like a lottery ticket.

A practical way to simplify:

- If your plan offers a target-date fund series, and the funds have reasonable costs, a target-date fund can be an excellent default. It's a diversified "set it and keep going" option that automatically shifts risk over time.
- If you prefer building your own mix, look for broad, low-cost index funds (U.S. stock, international stock, and a

bond fund). The names vary, but the idea is consistent: broad exposure, low expenses, no drama.

The most important number on the fund list isn't last year's return. It's the *expense ratio*—the annual fee charged by the fund, expressed as a percentage. A difference that looks tiny (0.80% vs. 0.08%) can matter a lot when applied every year to a growing balance. Fees compound in reverse.

If you're overwhelmed by choices, use this rule: pick one diversified fund you can explain in a sentence. "This is a target-date fund that holds a mix of stocks and bonds." Or: "This is a total stock market index fund." If you can't explain what you bought, you're relying on luck.

Fees beyond the funds: the costs you don't see unless you look

Many people assume their only cost is the fund expense ratio. Plans can also charge administrative fees or recordkeeping fees. You shouldn't panic about this—some plans are excellent, some are mediocre, and many are fine. The goal is visibility.

Your plan is required to disclose fees in participant communications, often including an annual comparative chart and periodic statements showing certain fees actually charged. Once you know where to look, it's not hard to find. The hard part is remembering that a fee you never notice is still real money.

If your plan's fees are high, that doesn't automatically mean you should avoid it, especially if the match is generous. Free money plus tax advantages can outweigh higher fees. But it does mean you should be more deliberate about getting in, capturing the match, and using the best low-cost options available inside the plan.

The mechanics that matter: paycheck timing, bonuses, and the ``oops" factor

Workplace plans run on payroll. That sounds mundane, but it has consequences.

If you contribute as a percentage of pay, your contributions automatically rise with raises and overtime, and you'll naturally contribute more in high-income years. If you contribute a flat dollar amount per paycheck, your saving is steadier, but you might miss out on capturing the match in a bonus-heavy year if the match is calculated per paycheck rather than annually (plans vary).

There's also the year-end problem: if you front-load contributions early in the year and hit the annual limit too soon, you may stop contributing—and depending on plan rules, you could miss match dollars later in the year. Some plans have a "true-up" feature that reconciles this. Some don't. If you're a high earner or an aggressive saver, it's worth asking HR or the plan provider one simple question: *Does the plan true up the match if I hit the limit early?*

That question can prevent an accidental self-own.

A workplace plan is also a behavior tool

There's a sneaky advantage to a 401(k) or 403(b) that has nothing to do with tax law: it makes saving automatic.

Automatic saving matters because willpower is seasonal. Some months you're motivated. Some months your car battery dies and your teenager needs braces and the dog eats something it shouldn't.

Payroll deferrals don't care about your month. They happen while you're busy living your life.

This is why the best workplace-plan strategy is often boring on purpose: set a contribution rate, pick a sensible investment, and keep showing up. The excitement belongs in your hobbies, not in your retirement account.

Actionable takeaway

Open your benefits portal (or call the plan provider) and do three things in one sitting:

1. Find your employer match formula and calculate the contribution percentage required to get the full match.
2. Set your contribution to at least that percentage starting with your next paycheck (even if you start smaller and schedule an automatic increase).
3. Confirm your vesting schedule for employer contributions and write down your next vesting milestone date.

Then pick a simple investment option you can live with—either a reasonable-cost target-date fund or a broad, low-cost index fund. The goal is not to impress anyone with fund selection. The goal is to turn on the plan's best features and keep them turned on long enough for them to matter.

5.3 Unlocking Catch-Up Contributions

The calendar turns, you hit 50, and suddenly the retirement system treats you differently—in a good way. It's one of the few moments in personal finance where aging unlocks something concrete, not just a new AARP envelope in the mail. Catch-up contributions are the tax code's quiet admission that the back half of a career has different math: you may finally have more income, more stability, and more urgency, all at the same time.

That urgency isn't a moral failing. It's simply what happens when the finish line stops being abstract. At 35, retirement is a foggy idea. At 52, you can almost hear it clearing its throat. Catch-up contributions turn that pressure into capacity: more room inside the same powerful, payroll-fed accounts you already have access to.

Catch-up contributions, in plain English

A catch-up contribution is exactly what it sounds like: an extra amount you're allowed to contribute to certain retirement accounts once you reach a certain age, on top of the regular annual limit.

The headline benefit is obvious—you can save more. The less obvious benefit is emotional: you get a fresh set of levers. When you're late to the game, the worst feeling is being trapped by small limits. Catch-up rules widen the doorway precisely when you're most motivated to walk through it.

For workplace plans, the most common catch-up threshold is age 50. The year you turn 50, the door opens (it's based on your age by the end of the calendar year, not the day you make the contribution). That means if you turn 50 anytime in 2026, you're treated as catch-up eligible for 2026.

The numbers that matter (and why the exact numbers aren't the point)

Contribution limits change over time, so treat the specific dollar amounts as a snapshot, not a tattoo. For 2026, the elective deferral limit for many workplace plans (401(k), 403(b), governmental 457(b), and the federal TSP) is $24,500. If you're age 50 or older, the standard catch-up contribution for many plans adds $8,000, for a combined employee limit of $32,500.

There's another wrinkle that's new enough to surprise even diligent savers: for certain plans, ages 60–63 may have access to a higher catch-up amount. For 2026, that higher catch-up is $11,250 in eligible plans.

Those numbers matter because they tell you what's possible. But the deeper point is this: after 50, the system allows you to compress decades of saving behavior into a shorter runway. You're not *supposed* to rely on it, but you *can* use it aggressively.

A compact way to see the 2026 workplace-plan landscape:

Workplace plan employee contribution limits (2026, many plans)	*Amount*
Elective deferral limit (most participants)	$24,500
Age 50+ standard catch-up (adds to limit)	$8,000
Total for age 50+ (elective deferral + catch-up)	$32,500
Ages 60–63 higher catch-up (eligible plans, 2026)	$11,250

If your reaction is, "There is no universe where I can contribute $32,500," that's fine. The limit isn't a commandment; it's a maximum. The question isn't whether you can hit the ceiling tomorrow. The question is whether you can climb higher than you're currently climbing, using every legal inch the ceiling gives you.

Why catch-up dollars punch above their weight

A dollar saved at 52 is not as powerful as a dollar saved at 22, purely in terms of time. That's not an insult; it's compound interest being honest. So how do you make late dollars matter?

You stack advantages. Catch-up contributions help you do that in three ways:

- They let you save more in accounts where growth is sheltered from annual tax drag.
- They often happen through payroll, which removes the monthly decision fatigue.
- They can increase your chance of capturing the full employer match, especially if you were under-contributing.

There's also a subtle psychological boost: catch-up contributions convert age into permission. Many people *want* to save more after 50 but feel guilty tightening the budget. The catch-up limit is a nudge from the system that says, "Yes, now. This is the moment."

A concrete case: Sheila turns 50 and buys herself options

Sheila is 49, works in operations for a mid-sized manufacturer, and has spent most of her adult life being the responsible one in the family. Two kids. A mortgage. A few years where life demanded cash more loudly than retirement demanded contributions.

She's contributing 6% to her 401(k) because that captures the full employer match. It's a great start, but it's not going to close the gap she can now see in her projections. The year she turns 50, she decides to treat catch-up like a structured pay raise for her future self.

Here's what she does instead of making a grand, dramatic leap:

- She keeps her 6% contribution in place (so the match stays locked in).
- She adds $150 per paycheck as an additional amount (a flat dollar catch-up layer).
- She schedules an automatic increase of $25 per paycheck every six months.

Nothing about that is flashy. It's intentionally unromantic. But it turns catch-up into a ramp instead of a cliff. And because the contributions are pre-tax (in her case), the increase feels smaller than the line item would suggest.

The real win is what happens a year later: Sheila doesn't feel like she's constantly deciding to be disciplined. The system is doing it for her. She can focus on living her life while her retirement contributions quietly become more ambitious in the background.

The catch-up gap: seeing acceleration instead of shame

People often talk about "being behind" in a way that creates panic or paralysis. Catch-up contributions offer a more useful frame: acceleration. When the rules allow you to save more after 50 than you could at 40, the curve can steepen. Not because markets got kinder, but because your savings rate got more powerful.

The next chart makes that acceleration visible. It compares the cumulative contribution capacity of a 40-year-old versus a 50-year-old over the next decade, using the 2026 workplace-plan limits as a simple baseline. Real life won't keep limits perfectly flat for ten years, but the shape of the advantage is the point: the gap is not a one-time bump; it's a widening wedge.

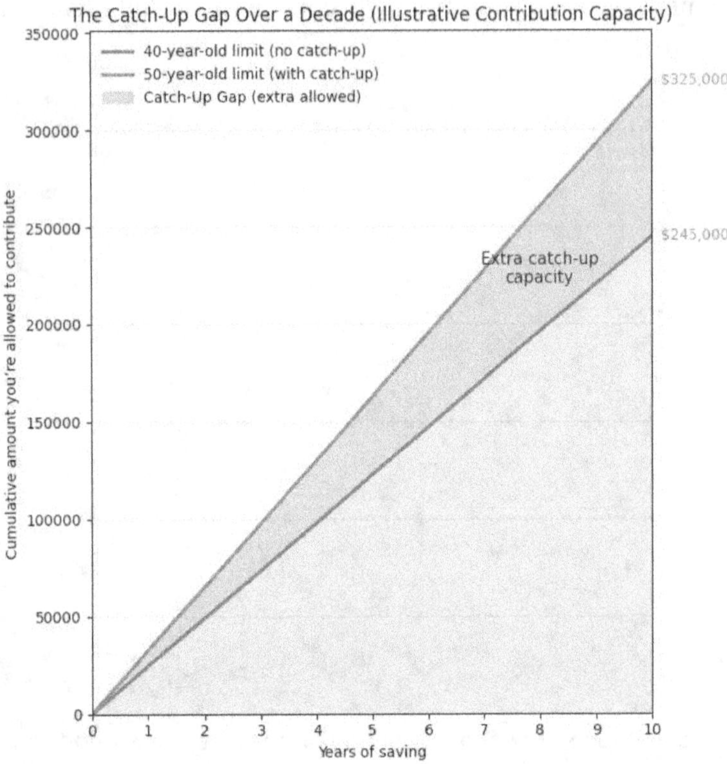

The orange wedge is the story: each year you're eligible for catch-up, the advantage stacks. That's why the best time to use catch-up contributions is rarely "someday." The wedge doesn't appear all at once; it grows by showing up consistently.

Ages 60--63: the late-career turbo button (if your plan supports it)

The 60–63 catch-up enhancement is the kind of policy that makes intuitive sense: if someone is close enough to retirement that their savings window is narrow, give them more room.

If you're in that age range (or approaching it), it's worth checking whether your plan is eligible and operational for the higher catch-up amount. Workplace retirement rules have been evolving, and implementation can vary. The practical move is simple: ask your plan provider or HR, *"What catch-up amounts does our plan allow for my age, and do I need to do anything special to use them?"*

This isn't busywork. It's the difference between believing you're maxing out and actually maxing out.

The Roth catch-up curveball: a reason to confirm, not a reason to freeze

Catch-up contributions come with a modern complication: for certain higher earners, catch-up contributions may be required to be made as Roth (after-tax) contributions, depending on the rules in effect and how a specific plan implements them. This is exactly the kind of detail that can scare people into doing nothing.

Don't give it that power.

Two grounding truths help:

- Even if some catch-up dollars must be Roth for you, that doesn't make them bad dollars. Roth contributions can be extremely valuable, especially for tax diversification.
- The right response is to confirm what your plan supports today and what it requires for your income level, not to abandon the whole idea.

If you hear coworkers swapping rumors about what's "allowed now," treat it like workplace folklore. Go to the source: your plan's actual rules, your payroll system's options, or the plan provider's participant line. You're not asking for trivia. You're asking how to route your money correctly.

How to afford catch-up contributions without pretending you'll become a monk

The dream version of catch-up is writing a huge check to your future. The real version is finding money in a life that is already busy and expensive.

The most reliable funding sources for catch-up contributions are not the ones that require heroics. They're the ones that come from life transitions:

- A car loan paid off (keep the payment, redirect it).
- A child finishing daycare (a cash-flow earthquake you can harness).
- A raise (commit a portion before it becomes lifestyle).
- A bonus (direct a set percentage to retirement before it hits checking).

The common theme is *capture the change*. If you wait until the money feels "extra," it won't. Lifestyle expands fast. Quietly redirecting new cash flow into the plan is how catch-up contributions become sustainable instead of aspirational.

The payroll trick that makes big contributions feel smaller

If you try to "save $8,000 this year" as one big goal, it feels heavy. If you translate it into payroll, it becomes a series of smaller, almost forgettable moves.

For someone paid biweekly (26 paychecks), an extra $8,000 per year is about:

$$\$8{,}000/26 \approx \$308 \text{ per paycheck.}$$

That number might still feel big, but now it's tangible. You can test it. You can start at $50 per paycheck, then $100, then $150. The plan doesn't require you to leap to the maximum. It rewards you for climbing.

And because traditional contributions reduce taxable income, the reduction in take-home pay is typically less than the contribution itself. That earlier "pre-tax dollars are cheaper" idea becomes especially practical here: catch-up contributions are often easiest to implement right where the taxes are being withheld.

The difference between ``maxing out'' and ``maxing out wisely''

There's a temptation among late savers to treat maxing out as the finish line: hit the limit, done. It's a great goal. But don't let the adrenaline of a bigger contribution distract you from the basic checks that keep the plan working well:

- Are you capturing the full match? (If not, fix that first.)

- Are your contributions set up so you don't accidentally miss match dollars later in the year due to front-loading and plan rules?
- Are you invested in something sensible rather than sitting in a stable-value or money market fund by accident?

None of these require you to become a retirement-plan scholar. They require you to be the kind of adult who looks under the hood occasionally.

Actionable takeaway

Open your payroll or benefits portal and set a specific catch-up target for the next 90 days, not the next decade. If you're turning 50 this year (or already there), do this:

- Convert an annual catch-up goal into a per-paycheck amount (even $25–$100 per paycheck counts).
- Turn on an automatic increase (for example, $25 more per paycheck starting in 3 months).
- Send one message to HR or the plan provider: *"I'm catch-up eligible. What is the maximum catch-up contribution our plan allows for my age in 2026, and are there any special Roth rules that apply to me?"*

Then let the calendar do what it does best: keep moving forward while your contribution rate quietly becomes the kind of number your future self will recognize as a turning point.

5.4 IRAs: Flexibility for Your Future

There's a particular kind of frustration that shows up in your 40s and 50s: you're finally willing to take retirement seriously, and then you discover your job's plan isn't as accommodating as your motivation. Maybe you're self-employed. Maybe your employer's 401(k) is fine but the investment menu feels like a vending machine with three stale options. Maybe you're maxing the match and still want more room. Or maybe you changed jobs enough times that you've learned to distrust anything that requires HR to explain it.

That's where the IRA earns its reputation as the Swiss Army knife of retirement accounts. An IRA (Individual Retirement Arrangement) is a retirement account you open yourself, separate from your employer. You pick the provider. You pick the investments. You decide whether each contribution is meant to lower today's tax bill (Traditional) or buy tax-free withdrawals later (Roth). You're not begging your payroll system for permission; you're steering.

The IRA is also where people get stuck, because it asks a deceptively hard question: *Do you want the tax break now, or later?* It's not a trick question. It's a bet on your own future.

Traditional vs. Roth: the simplest, most honest definition

A *Traditional IRA* is built around tax deferral. Depending on your income, filing status, and whether you're covered by a workplace retirement plan, your contribution may be fully deductible, partially deductible, or not deductible at all. If it's deductible, you may reduce taxable income today. The money generally grows tax-deferred, and withdrawals in retirement are generally taxed as ordinary income.

A *Roth IRA* is the opposite timing. Your contributions are not deductible. You pay taxes on those dollars now. If you follow the rules, qualified withdrawals later can be tax-free.

That's the clean version. The messy version is where the interesting decisions live: income limits, deductibility phase-outs, the reality that many people will have both types of money by retirement, and the fact that late savers are often balancing multiple financial goals at once.

The IRA decision isn't about picking the universally best account. It's about picking the best *next* dollar's tax treatment for your actual life.

The real question: what tax rate are you buying or selling?

When you choose between Traditional and Roth, you're doing a trade across time.

- With a Traditional contribution (if deductible), you *sell* your current marginal tax rate. You get a discount today, priced at today's marginal rate.
- With a Roth contribution, you *buy* your current tax rate. You pay it now to avoid paying tax later on qualified withdrawals.

So the decision reduces to a question that's both boring and profound: *Will your tax rate on withdrawals in retirement be higher or lower than your marginal rate today?*

That's why you hear the usual advice:

- If you're in a high bracket now and expect lower income in retirement, Traditional looks attractive.
- If you're in a lower bracket now and expect higher income later, Roth looks attractive.

But late savers rarely have a simple income story. They may have peak earnings now, but also large future required minimum distributions (RMDs) from existing pre-tax balances. They may plan to keep working part-time. They may inherit money. They may have a pension. They may want to retire early. They may be trying to keep Medicare premiums or taxes on Social Security benefits from rising later. The question isn't merely "high bracket or low bracket." It's: *What will your taxable income look like when you're actually living off your accounts?*

You don't need a perfect forecast. You need a plausible one.

The IRA's superpower: you control the provider and the investments

Workplace plans can be excellent, but they're often a closed ecosystem: a set menu of funds, plus whatever fees the plan negotiated (or failed to negotiate). An IRA is different. You can open it at a brokerage or mutual fund company that offers low-cost funds. You can build a simple, diversified portfolio with a handful of broad index funds. You can rebalance without calling HR.

That flexibility matters after 40 because every friction point becomes more expensive. If you only have a decade or two to compound, you don't want your plan to be a puzzle you avoid. You want it to be a tool you use.

IRAs also shine for people who don't have a workplace plan at all: gig workers, consultants, small business owners, or anyone between jobs. The IRA keeps your retirement saving from being held hostage by your employer.

The fine print that changes the answer: deductibility and income limits

Here's the part that trips people up: the Traditional IRA isn't automatically deductible for everyone. Your ability to deduct can depend on whether you (or your spouse) are covered by a workplace plan and on your income.

That means a late saver can innocently make a Traditional IRA contribution assuming it reduces taxes, only to learn at tax time that it doesn't. The contribution still belongs in the IRA; the issue is simply that it was *nondeductible*. Nondeductible Traditional contributions are allowed, but they create a record-keeping responsibility: you must track your *basis* (the after-tax amount you've already paid tax on), typically by filing Form 8606. Basis matters later because it determines what portion of distributions (or Roth conversions) should not be taxed again.

Roth IRAs have their own gatekeeping: eligibility depends on income. Many higher earners are phased out of direct Roth contributions. If you're over the limit, you don't get to "just do Roth" without planning.

The practical takeaway is not to memorize phase-out ranges. It's to respect that IRA contributions are not one-size-fits-all. Before you contribute for the year, confirm whether:

- a Traditional IRA contribution will be deductible for you, and
- you're eligible to contribute directly to a Roth IRA.

This is a five-minute check that can prevent a year of confusion.

A case study: two spouses, one paycheck, and a smarter IRA plan

Kevin is 55 and earns $135,000. His spouse, Tasha, is 53 and has stepped away from paid work for a few years to care for a parent. Kevin's employer offers a solid 401(k), and he's contributing enough to capture the full match, but their savings still feel thin for the timeline. They want an IRA strategy.

Their first assumption is common: "Since Tasha isn't working, she can't contribute." Not true. A *spousal IRA* allows a non-working spouse to contribute based on the working spouse's earned income, as long as they file a joint return and meet the other rules.

Now the real choice: Traditional or Roth?

Kevin's income is high enough that a Traditional IRA deduction may be limited or unavailable depending on plan coverage and income rules. If they contribute to a Traditional IRA and it's nondeductible, they *can* do it, but they've created basis tracking work (Form 8606) and future tax complexity. Meanwhile, their income may also be high enough to restrict direct Roth contributions.

Instead of guessing, they do two practical things:

- They check whether Kevin's Traditional IRA contribution would be deductible.
- They check whether they can contribute directly to Roth IRAs.

If Roth is available, Kevin and Tasha might decide that Roth contributions fit their priorities: they already have pre-tax exposure through the 401(k), and Roth IRAs would give them tax diversification. If Roth is not directly available, they might still be able to fund retirement through other routes, but the key is that they stop treating the IRA as a generic bucket and start treating it as a tax tool.

The emotional win in their story is subtle: Kevin stops thinking of retirement as a problem he has to solve alone. The spousal IRA turns their household into a two-engine savings system again.

Tax diversification: why many late savers benefit from having both types

Some people want a definitive answer: "Which is better, Traditional or Roth?" The honest answer is that many households are better off with a mix, because retirement is not a single tax event. It's a long series of withdrawals, year after year, with shifting rules, markets, and personal circumstances.

Having both pre-tax and Roth money can give you flexibility later:

- Need a lower taxable income year? Pull more from Roth.

- Comfortable staying within a lower bracket? Pull from Traditional up to a target and supplement with Roth.
- Want to manage how much of your Social Security benefits become taxable or how much taxable income shows up on your return? A Roth reserve can help.

Late savers often underestimate how valuable flexibility is. You can't predict tax law perfectly. But you can avoid concentrating all your retirement money in one tax treatment.

The IRA is a clean place to build that flexibility, especially if your workplace plan contributions are mostly pre-tax by default.

The withdrawal rules: the IRA is flexible, not consequence-free

IRAs are retirement accounts, not general savings accounts in disguise. If you withdraw from a Traditional IRA before age $59\frac{1}{2}$, you'll generally owe ordinary income tax and may owe an additional 10% penalty unless an exception applies. That doesn't mean IRAs are "locked away forever," but it does mean you should plan not to treat them as emergency funds.

Roth IRAs have their own rhythm. Contributions (the amounts you put in) can often be withdrawn more flexibly than earnings, but rules matter, and you don't want to discover your personal interpretation of Roth rules was optimistic.

The point for late savers is straightforward: build liquidity outside retirement accounts so you don't raid them. Raiding an IRA doesn't just create taxes; it collapses the compounding you were counting on.

Nondeductible Traditional IRA contributions: useful, but only if you respect the paperwork

There's a particular kind of tax mistake that's both common and expensive: making nondeductible Traditional IRA contributions and then forgetting about basis.

Basis is your after-tax "already paid" money inside the Traditional IRA. If you don't track it, you risk paying tax twice: once when you earned the money, and again when you withdraw it.

Form 8606 is the usual tool for tracking nondeductible contributions and basis. This is not glamorous. It is, however, the difference between an IRA strategy that quietly works and one that quietly leaks.

If you have ever made a nondeductible contribution in the past and you're not sure you filed Form 8606, that's not a reason to panic. It's a reason to gather your prior returns and confirm. Late savers don't need to be perfect; they need to be organized enough to protect themselves from avoidable double taxation.

The IRA contribution itself is only half the decision

Many people do the hard part (they contribute) and then accidentally stop at the wrong moment. Money can land in an IRA settlement fund or money market fund and sit there, earning very little, because no one completed the second step: investing it.

This is one of the most painful unforced errors in retirement planning because it looks responsible on paper: "I contributed to my IRA." But the growth engine never turns on.

A simple safeguard is to set an automatic investment instruction when you open the account, or to schedule a recurring purchase into a diversified fund. If you prefer hands-off, choose a single diversified fund you understand (for example, a target-date index fund series) and direct contributions there automatically.

The goal is not to make investing exciting. The goal is to make it inevitable.

How to choose when you're genuinely unsure

If you're torn between Traditional and Roth and you don't want to pay for a full financial plan, use a practical framework based on three questions:

1. *Is a Traditional IRA contribution deductible for me?* If not, the comparison changes. Nondeductible Traditional contributions can still be useful, but they're not the same as a deductible contribution.
2. *Am I eligible for a direct Roth IRA contribution?* If yes, Roth becomes a straightforward choice for many households seeking tax diversification.
3. *Do I expect my taxable income in retirement to be meaningfully lower than today?* If yes, deductible Traditional contributions may be especially attractive.

This isn't a quiz with one correct answer. It's a way to avoid the two biggest mistakes: contributing based on assumptions ("Traditional is always deductible" or "Roth is always allowed"), and letting indecision stop you from contributing at all.

Actionable takeaway

Before you make your next IRA contribution, do one small, high-leverage prep step: pull up last year's tax return (or your most recent paystub if that's easier) and write down your approximate marginal federal bracket. Then, in the same sitting, confirm two facts using your tax software, a reputable eligibility calculator, or a quick call to a tax pro:

- whether a Traditional IRA contribution would be deductible for you, and
- whether you're eligible to contribute directly to a Roth IRA.

Once you have those answers, set up an automatic monthly contribution to the IRA and an automatic investment into a diversified fund, so the account doesn't become a parking lot. The best IRA strategy isn't the one that sounds clever at a dinner party; it's the one that reliably moves money from your paycheck into invested assets every month, without requiring you to feel brave or inspired.

5.5 Consolidating Your Financial History

Most people don't have a retirement *plan* so much as a retirement *trail*. A 401(k) from the job you loved. A 403(b) from the job you tolerated. A little IRA you opened one year when you were feeling ambitious. Maybe a pension statement that shows up like a postcard from another life. If you're over 40 and you've changed jobs a few times, your retirement money can start to look like a sock drawer: technically organized, emotionally chaotic.

The cost of that chaos isn't only aesthetic. Scattered accounts are easier to forget, harder to invest consistently, and more likely to carry unnecessary fees. They also make it harder to answer basic questions that should be simple by now: *How much do I have? What is it invested in? Am I taking the right amount of risk?*

Consolidation is not a financial flex. It's a stress-reduction strategy that can quietly improve outcomes by making it easier to do the boring, right things repeatedly.

The hidden tax of ``I'll deal with it later"

Old accounts are sticky because dealing with them rarely feels urgent. You're busy starting a new job, learning a new system, surviving a new commute, or just enjoying the relief of leaving a bad situation. The old 401(k) sends a statement once in a while. The balance doesn't look terrible. You think, *I'll roll it over when I have time.*

Then years pass.

During those years, three things can happen without your permission:

- You lose track of the account entirely after moving addresses or changing emails.

- The money sits in an overly conservative default investment (sometimes cash-like) after a job change, quietly underperforming.
- Fees that seemed small when the account was active become annoying when the account is just sitting there, unattended.

None of this is dramatic enough to trigger action, which is exactly why it's so costly. It's not one big mistake; it's a long stretch of unoptimized autopilot.

First, locate everything: you can't consolidate what you can't find

Before you move money, you need an inventory that's more reliable than memory. The safest assumption is that you're forgetting something—not because you're irresponsible, but because adulthood is noisy.

A practical inventory includes:

- current employer plan(s): 401(k), 403(b), 457(b), TSP
- old employer plans, even if the balance is small
- IRAs (Traditional, Roth, rollover IRAs)
- any pension or cash-balance plan statements
- HSAs (not a retirement account, but often part of the long-term picture)

The easiest way to jog your memory is to use your work history as a checklist. Start with your current employer and work backward. For each job, ask: *Did I have a retirement plan there? Did I contribute? Did the employer contribute?*

Then use paper trails. Old W-2s are especially useful because they often show retirement plan participation indicators and the employer name and address from that year. Old emails can help too: search your inbox for "401(k)," "403(b)," "retirement plan," "recordkeeper," or the names of common providers.

If you find an account but can't access it, don't assume the money is gone. Accounts can be transferred between record-keepers when employers change providers. Your job is simply to keep following the thread.

Your four basic options when you leave a job

When you leave an employer, you typically have four broad choices for that employer's plan money (assuming it's vested and the plan allows the option):

1. Leave it in the old plan.
2. Roll it to your new employer's plan.
3. Roll it to an IRA.
4. Cash it out (the option that often looks tempting and usually hurts the most).

These aren't equal choices. They're tradeoffs.

Leaving money in the old plan can be fine if the plan is low-cost and offers great investment options. Rolling to a new employer plan can simplify life and keep everything under one roof, but some plans have limited investment menus or higher fees. Rolling to an IRA often gives you the broadest investment choice and clear visibility, but it may change certain protections and may complicate things if you're considering certain tax strategies later. Cashing out is almost always the option you choose because life is on fire, and it tends to come with taxes and penalties that turn a retirement account into an expensive emergency loan.

Consolidation doesn't mean "IRA is always best" or "new 401(k) is always best." It means making a deliberate choice rather than defaulting into five separate accounts you never look at.

The rollover: the move that must be done the boring way

A rollover is simply moving money from one retirement account to another. The reason rollovers are worth respecting is that the tax consequences depend heavily on *how* you move the money.

There are two main styles:

- *Direct rollover*: the money moves from the old plan directly to the new plan or IRA custodian. You never take possession of the funds.
- *60-day rollover*: the distribution is paid to you, and you have 60 days to redeposit it into an IRA or another eligible retirement account.

If you remember one thing, make it this: *direct rollovers are the boring, safe default.*

Here's why. When a distribution is paid to you (even if you intend to roll it over), plans are generally required to withhold 20% for federal taxes. That means if you try to roll over the full amount, you have to come up with that missing 20% out of pocket to complete the rollover. Many people can't or don't, which turns part of the distribution into a taxable event. And if you miss the 60-day deadline, the distribution can become taxable income, and you may owe penalties if you're under $59\frac{1}{2}$.

The direct rollover avoids the withholding trap and the 60-day clock. It's not exciting. It's simply cleaner.

A concrete case: Danielle's ``withholding gap" surprise

Danielle is 46 and switched employers after a long stint at a nonprofit. She had about $62,000 in her old 403(b). During the job transition, she decided to roll it into an IRA so she could simplify and choose lower-cost investments. She requested a distribution, expecting to deposit it into the IRA once it arrived.

The check showed up for about $49,600, not $62,000. The plan had withheld roughly 20% for federal taxes. Danielle was confused and annoyed, but she still wanted to complete the rollover. To roll over the full $62,000, she would have needed to deposit the $49,600 *plus* an additional $12,400 of her own money into the IRA within 60 days. She didn't have $12,400 sitting around.

So she rolled over only what she received. The withheld amount became, in effect, a forced distribution. At tax time, she owed taxes on that amount as income. She also owed a penalty because she was under $59\frac{1}{2}$ and no exception applied. Danielle didn't do anything reckless; she just used the wrong method.

The lesson is not "Danielle should have known." The lesson is that rollovers are paperwork-heavy for a reason: the tax system is trying to keep you from treating retirement accounts like checking accounts. The safest way to consolidate is to never let the money touch your hands.

The sharp edges: not everything is eligible for rollover

Another reason to be methodical is that some distributions are not eligible to roll over. Required minimum distributions (RMDs), for example, generally can't be rolled over. Certain hardship distributions also may not be eligible. If you're older or if the account is in a phase where distributions are happening, you need to confirm what portion (if any) is rollover-eligible before initiating the move.

This is especially relevant for late savers because consolidation often happens later than ideal. People start looking seriously at their accounts in their 50s and 60s, precisely when distribution rules begin to matter more. A rollover that's straightforward at 45 can be more nuanced at 73.

The practical fix is simple: when you call the old plan, use the exact phrase *"direct rollover of the full eligible rollover distribution"*. You're signaling that you want the clean version, and you're prompting the representative to identify any portion that can't be rolled over.

The one-rollover-per-year rule: the mistake that sounds fictional until it happens

IRAs have a rule that surprises people: in many cases, you can only do one IRA-to-IRA rollover in any 12-month period (aggregating all of your IRAs). This rule generally applies when you take a distribution and then redeposit it (the 60-day rollover style), not when you do a direct trustee-to-trustee transfer.

This is one more reason direct transfers are the adult choice. If you move money between IRAs using trustee-to-trustee transfers (where the custodians handle the movement), you reduce the chance of stumbling into a rule you didn't know existed.

If you're consolidating multiple old accounts into one IRA, you may be tempted to "collect checks" and deposit them. Resist that temptation. Let institutions move the money directly whenever possible.

So where should the money go? Use a decision that matches your life

Consolidation should make your life easier. It should also respect the role your accounts play.

A useful way to decide between rolling old plan money into a new employer plan versus into an IRA is to compare along a few dimensions:

- *Fees and investment options*: Is the old plan expensive? Is the new plan better? Would an IRA offer materially lower costs or better fund choices?

- *Simplicity*: Will consolidating into the new plan reduce the number of accounts you manage? Or will an IRA be the simplest "one dashboard" solution?
- *Withdrawal flexibility and rules*: Employer plans and IRAs can differ in loan options, distribution rules while still employed, and administrative processes.
- *Behavior*: Where are you more likely to leave the money invested and alone? Some people tinker more in an IRA because it's accessible; others are more disciplined when the money is tucked into a workplace plan.

A late saver's best consolidation target is often the place that makes contributions and maintenance easiest, not the place with the theoretically optimal feature set.

The best account is the one you will actually use well.

How to do a clean rollover: a step-by-step that avoids drama

Once you've decided where the money should land, the actual process can be made remarkably predictable:

1. *Open the receiving account first.* If you're rolling into an IRA, open it at the custodian you want and confirm the account type (Traditional to Traditional, Roth to Roth, etc.).
2. *Call the receiving institution and ask for rollover instructions.* Many custodians will generate a letter of acceptance or provide specific wording for checks.
3. *Initiate a direct rollover from the old plan.* Specify that the check should be made payable to the receiving custodian

for the benefit of (FBO) you, or that the funds should be sent electronically if available.

4. *Track the money like you would track a passport.* Get confirmation numbers, expected timelines, and a contact person or department.
5. *Reinvest promptly on arrival.* Decide in advance what the money will be invested in so it doesn't sit in cash for weeks.

If this sounds like a lot, remember what you're doing: moving a significant pool of money through the financial system without taxes being triggered. A little paperwork is the price of keeping the IRS out of your compounding.

The ``financial junk drawer" problem: small accounts that steal attention

One of the quiet benefits of consolidation is removing the mental clutter of small, orphan accounts. A $3,200 401(k) from a job you had in 2009 doesn't feel like a big deal. But it creates a new login, a new statement, a new beneficiary designation to keep current, and a new place where money might be invested poorly by default.

Late savers don't lack intelligence. They lack spare cognitive bandwidth.

When you consolidate, you're not just optimizing fees. You're buying back attention. And attention is the resource that makes it possible to do higher-value planning later, like setting contribution targets, choosing a sensible asset allocation, and staying the course when markets are unpleasant.

A note on beneficiary designations: the paperwork that matters more than the investments

Consolidation often triggers a valuable housekeeping step: updating beneficiaries. Retirement accounts pass by beneficiary designation, not by whatever your will says in many cases. That means an old account you forgot about can still list an ex-spouse or a parent as beneficiary simply because you never updated the form.

This is not a theoretical problem. It's a real, messy, heartbreaking problem.

When you consolidate, you have a natural moment to confirm:

- primary beneficiary
- contingent beneficiary
- whether the designations reflect your current family reality

This is one of those five-minute tasks that can prevent a future legal and emotional disaster.

Actionable takeaway

Schedule a 45-minute "account hunt" this week and treat it like an appointment. Create a single list (a spreadsheet or a note) with three columns: *Provider, Account type, Approximate balance*. Work backward through your employers and fill in the list until you can account for every retirement dollar you've ever earned.

Then pick one old account—the one with the highest fees, the worst investment options, or the greatest risk of being forgotten—and initiate a *direct rollover* into your chosen consolidation destination. Do not request a check made out to you. Do not accept 20% withholding. Do not start a 60-day clock you didn't need.

When the money lands, invest it within 48 hours in a simple, diversified allocation you can explain in a sentence, and update the beneficiary designation before you close the browser tab.

5.6 Tax-Efficient Investing Habits

People love to argue about investments the way sports fans argue about teams: small differences, loud opinions, lots of cherry-picked stats. But taxes are quieter and far more faithful. They show up every year. They don't care whether you picked the "right" fund. They simply take their cut of whatever your account happens to produce.

Tax-efficient investing habits aren't about being clever. They're about refusing to let your portfolio spring tiny leaks for decades. After 40, that matters more than it sounds. You have less time to recover from avoidable friction, and you're more likely to have meaningful dollars in multiple account types: a workplace plan, maybe a Roth, maybe a taxable brokerage account, maybe an HSA. The question stops being "What should I invest in?" and becomes "Where should I hold each investment so the IRS bothers it the least?"

That question has a name: *asset location*. It's not asset *allocation* (how much you hold in stocks vs. bonds). It's asset *location* (which account holds which assets). Same ingredients, different shelves.

The tax personality of your investments

Every investment has a tax personality. Some are polite and barely cause taxable events. Others are loud roommates who leave messes in the kitchen every single day.

In a taxable brokerage account, taxes tend to come from three main places:

- *Interest*: Usually taxed at ordinary income rates. Bonds and cash-like investments are the usual culprits.
- *Dividends*: Some dividends may be taxed at preferential rates (qualified dividends) and some at ordinary income rates (nonqualified dividends). Either way, dividends create taxable income even if you reinvest them.
- *Capital gains*: If you sell for a gain, you owe tax. Less obviously, mutual funds can distribute capital gains when the fund trades internally, and that distribution can be taxable to you even if you didn't sell your fund shares.

This is the beating heart of "tax drag" in real life: you can be a disciplined long-term investor and still get a tax bill just for owning the investment in the wrong place.

In tax-advantaged accounts (Traditional 401(k)/IRA, Roth, etc.), the same investment activity generally doesn't create annual taxes. That doesn't make taxes disappear forever, but it does change the timing and, in Roth accounts, can potentially eliminate taxes on qualified withdrawals.

So asset location is essentially matchmaking: you pair tax-inefficient assets with tax-sheltered accounts, and you reserve taxable accounts for assets that behave well under taxation.

The three-account mental model that makes this easy

Most households with meaningful savings end up with some version of three "buckets":

- *Taxable*: brokerage accounts, joint accounts, individual accounts.
- *Tax-deferred*: Traditional 401(k), 403(b), Traditional IRA. You may get a tax break now; withdrawals are generally taxed later.
- *Tax-free*: Roth IRA, Roth 401(k) (and sometimes HSAs if used strategically for qualified medical expenses).

Asset location asks: what belongs where?

A simple heuristic (not a law, but a strong starting point) is:

- Put *tax-inefficient* assets (often bond funds, REIT funds, actively managed strategies with lots of turnover) in tax-deferred or tax-free accounts.
- Put *tax-efficient* assets (often broad stock index funds, ETFs with low turnover, stocks you plan to hold for a long time) in taxable accounts.

The goal is not perfection. The goal is to put your loudest tax roommates in the rooms where taxes don't complain every year.

Why bonds often belong in retirement accounts

Bonds are the classic example. Many bond funds pay interest that is taxed at ordinary income rates in a taxable account. That means you can end up paying relatively high taxes on relatively modest returns—a depressing combination.

If you hold a bond fund inside a Traditional 401(k) or IRA, you generally avoid annual taxation on that interest. Taxes are deferred until withdrawal. If you hold it inside a Roth, qualified withdrawals can be tax-free. Either way, you've stopped the yearly drip of taxes that reduces compounding.

This is not a statement about whether bonds are good or bad. Bonds can be essential for risk management, especially as retirement approaches. Asset location doesn't change whether you should own bonds. It changes where you should put them.

Why broad stock index funds often work well in taxable

Broad stock index funds and many ETFs can be relatively tax-efficient because they tend to have lower turnover (fewer internal trades). That can reduce capital gains distributions. Dividends still happen, but qualified dividends may receive preferential tax treatment, and the real magic of taxable stock investing is that you can control when you realize capital gains by choosing when to sell.

That control is powerful. It's the difference between taxes being a yearly fee and taxes being something you strategically trigger when it benefits you.

Taxable accounts also offer the possibility of *tax-loss harvesting*: selling an investment at a loss to offset gains (and potentially some ordinary income, within limits), then reinvesting in a similar (not substantially identical) investment to maintain market exposure. This isn't necessary for everyone, and it can be overdone, but it's a tool that only exists in taxable accounts.

The trap: treating asset location like a religion

Asset location can become a rabbit hole. Some people turn it into an obsession: spreadsheets, optimization, constant tinkering. That misses the point.

The real value of asset location comes from a few big, stable decisions:

- Don't put highly taxed income-producing assets in taxable if you have room in sheltered accounts.
- Don't waste sheltered space on assets that were already tax-efficient in taxable.
- Keep the whole portfolio's risk level sensible; taxes are not worth taking the wrong risk.

After 40, consistency beats cleverness. A slightly suboptimal setup that you stick with is better than a theoretically optimal setup you keep breaking because it's too complicated.

A concrete case: Paul and Mina stop paying ``voluntary" taxes

Paul (54) and Mina (52) have done a lot right. They both contribute enough to get their employer matches. They have a Roth IRA that they fund some years. They also have a taxable brokerage account they've been building for flexibility.

Their allocation across all accounts is roughly 70% stocks / 30% bonds, which fits their risk tolerance. But when they look closely, they realize where those assets live is backwards:

- In taxable: a bond fund and a high-dividend stock fund (because someone told them dividends are "income").
- In tax-deferred (401(k)s): mostly stock funds, because those were the default choices.

Their taxable account is throwing off interest and dividends every year, creating a steady tax bill. Meanwhile, their 401(k)s are holding the assets that could have been relatively tax-friendly in taxable.

They don't change their overall 70%/30% allocation. They simply swap *locations*:

- Bonds move into the 401(k)s.
- Broad stock index exposure shifts toward taxable.

They execute the change mostly through rebalancing in the retirement accounts (no tax consequences inside the 401(k)) and by directing new taxable contributions toward the stock index fund they want long-term. They avoid unnecessary selling in taxable that would create gains.

A year later, their tax bill from the taxable account is noticeably lower. More importantly, their portfolio feels cleaner. The accounts now have roles: retirement accounts shelter the tax-inefficient pieces; taxable accounts hold the patient, long-term growth engines.

Paul's reaction is the most telling: "We weren't paying taxes because we were rich. We were paying taxes because we were sloppy."

The wash sale rule: the one tax rule that punishes good intentions

Tax-loss harvesting can be useful, but it comes with a rule that is both simple and surprisingly easy to violate: the *wash sale* rule.

In plain language, if you sell an investment at a loss and buy the same or a "substantially identical" investment within 30 days before or after the sale (a 61-day window centered on the sale date), the loss can be disallowed for current tax purposes. Instead of getting the tax benefit now, the disallowed loss is generally added to the basis of the replacement shares.

The part that catches people after 40 is that their investing is spread across multiple accounts. You might sell an S&P 500 ETF in taxable for a loss, then your 401(k) automatically buys an S&P 500 index fund through payroll contributions a week later. Congratulations: you may have created a wash sale without intending to.

This is not a reason to fear taxable investing. It's a reason to coordinate. If you tax-loss harvest, do it with awareness of what's happening in your other accounts during the 30-day windows.

A conservative habit if you want to harvest losses: when you sell for a loss in taxable, replace it with something similar but not substantially identical (for example, a different broad U.S. equity fund that tracks a different index). Or pause automatic purchases of the same holding across accounts during the wash-sale window. You're not trying to outsmart the IRS. You're trying to avoid stepping on a rake.

High income and the NIIT: when the tax bite gets sharper

For higher-income households, there can be an extra layer of tax drag: the Net Investment Income Tax (NIIT), a 3.8% surtax that can apply to certain investment income above certain income thresholds. Not everyone pays it, but if you do, it raises the cost of taxable dividends, interest, and capital gains.

This is one reason asset location becomes more valuable as income rises. If taxable investment income is being taxed at higher effective rates, the benefit of sheltering tax-inefficient assets in retirement accounts is amplified.

This is also where the "I'll just invest everything in taxable for flexibility" approach can become quietly expensive. Flexibility is valuable. But when the tax rate on that flexibility creeps up, you want to be intentional about what income you're generating in taxable.

The overlooked habit: put your best growers where the IRS can't touch them

There's a second school of thought in asset location that can also make sense: put the highest expected growth assets in Roth accounts, because Roth growth can be the most valuable growth. The math is intuitive. A Roth dollar is already "after-tax"; if it grows dramatically and can later be withdrawn tax-free (under the rules), you've protected a larger future amount from taxation.

This doesn't mean you should jam all stocks into Roth and all bonds into Traditional without thinking. It means the Roth is precious real estate. Many late savers accidentally waste Roth space on cash or overly conservative allocations because they're nervous. Nervous is understandable. But if you're going to take risk somewhere, taking it inside the account with the best tax treatment can be rational.

The key is matching risk to your overall plan, not to one account. Your portfolio is one portfolio, even if it's scattered across accounts.

Rebalancing: the cleanest place to do it

Asset location also changes where you should rebalance. Rebalancing is the act of bringing your portfolio back to its target allocation (for example, back to 70/30 after stocks rise and you drift to 78/22). In taxable accounts, rebalancing by selling can create capital gains. In retirement accounts, you can often rebalance without current tax consequences.

A tax-efficient habit is to do most of your rebalancing in tax-advantaged accounts when possible, and to use new contributions in taxable accounts to steer allocation gradually rather than forcing big taxable sales.

This is one of those habits that feels small and ends up big: you keep your risk in line without generating avoidable taxes.

A practical ``good enough" asset location blueprint

If you want a simple starting point that fits many late savers, try this mindset:

- *Taxable account:* broad stock index funds / ETFs you expect to hold for a long time; keep turnover low.
- *Traditional (401(k)/IRA):* bond funds and other income-heavy assets; use this space for rebalancing.
- *Roth:* a diversified allocation with an eye toward long-term growth; avoid leaving it in cash by accident.

Then adapt it to your real constraints. Maybe your 401(k) has an excellent low-cost stock index fund and a mediocre bond fund. Fine. Use the best available options. Asset location is not about purity; it's about reducing unnecessary taxes with the tools you actually have.

Actionable takeaway

Open a note and list every account you have under three headers: *Taxable, Tax-deferred, Tax-free*. Next to each account, write the two or three largest holdings and whether each holding throws off a lot of taxable income (interest-heavy, high turnover, dividend-heavy) or is relatively tax-friendly.

Then make one targeted change that reduces tax drag without triggering a tax bill:

- Move bond exposure *inside* a 401(k)/IRA by changing future contributions there, and
- direct future taxable contributions toward a broad, low-turnover stock index fund.

If you do nothing else, that one habit—putting the tax-loud assets under a tax-quiet roof—can keep more of your portfolio's return in your pocket year after year, which is exactly what late-stage compounding needs: fewer leaks and more time spent actually growing.

Chapter 6

The Finish Line: Planning Your Retirement

Retirement isn't a cliff you simply fall off; it is a house you must actively build. For years, you have likely heard vague rules of thumb like 'save $1 million' or 'replace 80% of your income,' but generic advice fails when the runway gets shorter. In this chapter, we stop guessing and start measuring. We will calculate exactly what your future life costs, how to bridge the intimidating healthcare divide, and how to treat Social Security like the superpower it is. It is time to turn 'I hope I have enough' into 'I know I have a plan'.

6.1 Estimating Your Real Retirement Number

Retirement math gets treated like a fortune cookie: *"You'll need 80% of your income."* It sounds crisp. It sounds responsible. It also has the same problem as all one-size-fits-all advice: it assumes your life is average, your spending is smooth, and the future is a photocopy of your present.

But real retirement doesn't behave like that. Some costs evaporate the month you stop working. Others show up like they've been waiting in the wings. And the biggest mistake people over 40 make isn't that they save too little—it's that they aim at a number that has nothing to do with the life they actually plan to live.

A "real retirement number" is not a single scary target that floats in the distance. It's a price tag on a lifestyle. And like any price tag, it's built from line items.

6.2 Why the 80% rule keeps missing the point

The 80% rule is a shorthand for something sensible: many retirees spend less than they did while working. They're no longer saving for retirement, payroll taxes may be lower, commuting goes away, and the daily friction costs of work (lunches out, last-minute convenience spending, professional clothes, the "I deserve this" purchases after a hard week) often shrink.

The trouble is that income is not spending. Income is a *funding source*. Spending is the *goal*. If you're making $120,000 and spending $80,000, you don't need 80% of your income; you need something that reliably funds $80,000 (plus taxes and healthcare, plus surprises). If you're making $120,000 and spending $115,000 because you're supporting a parent, paying private school, or carrying a mortgage that ate your margin for a decade, 80% is a fantasy.

Also: 80% assumes retirement is flat. It isn't. People tend to spend in phases, not a straight line, and those phases can change your number more than any market forecast.

6.3 The three-phase lens: Go-Go, Slow-Go, No-Go

A useful way to stop thinking about retirement as a single blob of time is the *Go-Go, Slow-Go, No-Go* framework:

- *Go-Go years*: you're healthy enough to travel, try hobbies, say yes to things. Spending is often higher here, not lower.
- *Slow-Go years*: you still do plenty, but big trips and big days out taper. Spending often moderates.
- *No-Go years*: activity narrows. Some categories fall (travel, restaurants). Others rise (healthcare, help at home, transportation services).

This isn't a prediction about your personality. It's a reminder that retirement spending is not "inflation-adjusted current spending forever." It's *a moving mix*.

If you build your retirement number as if every year will look like your most expensive vacation year, you'll over-save or work longer than necessary. If you build it as if every year will look like your most frugal month, you're setting yourself up for anxiety-driven decisions later—the kind where you cancel plans that would have made your early retirement worth it.

6.4 Start with a bottom-up budget (yes, even if you hate budgeting)

There are two kinds of people: those who enjoy spreadsheets, and those who would rather clean gutters than categorize expenses. Either way, the bottom-up approach is the adult version of retirement planning because it forces clarity.

Here's the trick that makes it tolerable: you're not building *a perfect budget*. You're building *a credible draft*. The goal is to replace a myth (80%) with a map (your line items).

Begin with your current spending, but treat it like a closet clean-out:

- *Keep*: expenses that will still exist (utilities, groceries, insurance, property taxes, basic transportation, phone/internet).
- *Toss*: expenses that vanish (retirement contributions, commuting costs, payroll taxes tied to work, work-related child care, certain professional costs).
- *Alter*: expenses that change shape (housing, healthcare, travel, hobbies, gifting, supporting adult kids, dining out).

Then build three versions of the retirement budget: Go-Go, Slow-Go, No-Go. Not three totally different lives—just three different mixes.

A practical way to do this without losing your mind:

1. Make a list of your spending categories (12–20 is plenty).

2. Put your current annual spending next to each.
3. Create three columns labeled Go-Go, Slow-Go, No-Go.
4. Adjust only the categories that truly change.

You'll notice something immediately: the retirement number isn't a single number. It's a range, with early retirement often costing more than people expect.

6.5 The expenses that ``disappear'' (and the ones that don't)

Some line items are genuinely work-powered. Removing them can create surprising room.

Commuting is the obvious one: gas, parking, tolls, transit passes, car wear-and-tear, the second car "we need for the schedule." Many households don't just save commuting costs; they also change their entire transportation footprint.

Saving for retirement disappears too, and that's not a small thing. If you've been putting $15,000–$30,000 per year into a 401(k) and IRA, your spending can stay the same while your income need drops. This is why some people can retire on far less income than they earned.

But other expenses are sticky.

Housing is the big one. People talk about paying off the mortgage like it's a magic trick, and it is helpful—but a paid-off house still comes with property taxes, insurance, repairs, and the occasional surprise that costs the same as a used car. Downsizing can lower costs, but it can also *raise* them in the short term if you're buying into a more expensive area, paying HOA fees, or moving in a high-rate environment.

Family support is another sticky one. Plenty of people in their 40s and 50s are squeezed between adult kids and aging parents. Retirement doesn't automatically end that. Sometimes it increases it, because you become the flexible one.

Health-related costs are the wild card. Even in a calm year, healthcare spending in retirement tends to be a larger share of the budget than during working years. In a rough year, it can dominate the plan.

6.6 A concrete case: Maria and Devin, age 47 and 50

Maria is 47 and works in HR. Devin is 50 and runs a small contracting business. Combined gross income: about $165,000. They're classic "we should be doing better" earners: good salaries, busy lives, and retirement feels close enough to be loud but far enough to procrastinate.

Their current spending (rounded) looks like this:

- Housing (mortgage, taxes, insurance, basic repairs): $42,000
- Utilities/internet/phones: $6,600
- Groceries/household: $14,400
- Transportation (two cars, insurance, gas, maintenance): $13,800
- Dining/entertainment: $9,600

- Travel: $6,000
- Insurance (life/disability umbrella): $3,600
- Medical out-of-pocket (not premiums): $2,400
- Gifts/family support: $4,800
- Miscellaneous (subscriptions, clothes, everything else): $8,400

Total: about $112,000 in spending.

Here's what the 80% rule says. Eighty percent of $165,000 is $132,000. That would make them think they need *more income in retirement than they spend now,* which feels impossible and leads to one of two reactions: panic or avoidance.

Now watch what happens when they build a bottom-up Go-Go/Slow-Go/No-Go budget.

First, they identify what actually changes:

Mortgage payoff: they'll be mortgage-free at 63 if they stay on schedule. That reduces the housing line later, but not to zero.

Transportation: they expect to go down to one car when Devin stops working full time.

Travel: they want to do more early—two bigger trips a year for about 8 years.

Dining/entertainment: they expect a small increase early (more weekday flexibility, more social time), then a decrease later.

They build their Go-Go budget for ages 63–72 (their rough target window):

- Housing: $28,000 (no mortgage, but taxes/insurance/repairs remain)
- Utilities/internet/phones: $7,200
- Groceries/household: $15,600
- Transportation: $9,000 (one car + more local driving)
- Dining/entertainment: $11,400
- Travel: $14,000
- Insurance: $2,400 (life/disability drop; keep umbrella)
- Medical out-of-pocket: $4,000 (higher than today)
- Gifts/family support: $6,000 (they assume this doesn't magically vanish)
- Miscellaneous: $9,600

Go-Go total: about $107,200.

Their Slow-Go budget trims travel and a few lifestyle categories: Slow-Go total: about $92,000.

Their No-Go budget drops travel significantly but increases medical and home support: No-Go total: about $96,000.

Two important truths hit them at once:

1. Their retirement spending is not a straight line down. It dips, then shifts.
2. Their Go-Go number is *close* to their current spending, not 80% of income.

That changes their planning posture completely. Instead of aiming at $132,000 of income forever, they're aiming to fund roughly $90,000–$110,000 depending on the phase—and they can see *why*.

6.7 Convert spending into a portfolio target (without pretending the future is certain)

Once you have an annual spending target, you need a way to translate it into "how much money do we need invested?" This is where people often reach for a single magic number again: *the 4% rule.*

In plain English, the 4% rule is a rule of thumb that says: if you start retirement by withdrawing about 4% of your portfolio in the first year (and then increase that dollar amount with inflation), a diversified portfolio has historically had a decent chance of lasting 30 years.

It's not a guarantee. It's a starting point. And for someone starting later, it's best used as a *range* rather than a commandment—because the risk that matters most is not average return. It's bad timing. A market drop early in retirement can do outsized damage when you're withdrawing.

A practical approach:

- Use 4.0% as an *optimistic* baseline.
- Also run 3.5% as a *more conservative* baseline (especially if retiring in your early 60s, or if you want higher confidence).

If Maria and Devin want to fund $100,000 per year from investments (after accounting for any guaranteed income like Social Security later), here's what those two baselines imply:

- At 4.0%: $\$\frac{100{,}000}{0.04} = 2{,}500{,}000$ ($2.5 million)
- At 3.5%: $\$\frac{100{,}000}{0.035} \approx 2{,}857{,}143$ ($2.86 million)

That looks big until you remember: (1) they may not need to fund the entire amount from investments if Social Security covers a meaningful share, and (2) their spending is phased—they may choose a higher withdrawal rate in the Go-Go years and lower later, or plan travel as a "bucket" rather than a permanent expense.

The real win is psychological: the number is no longer a mysterious mountain. It's just annual spending divided by a planning rate.

6.8 Make room for taxes (the stealth expense)

Taxes are the most common missing line item in retirement budgets. People assume taxes drop because their income drops. Sometimes true. Sometimes not.

Retirement income can be taxed in messy ways because it often comes from multiple sources:

- taxable accounts (capital gains and dividends),
- tax-deferred accounts like traditional 401(k)s/IRAs (ordinary income),
- tax-free Roth accounts (qualified withdrawals),

- Social Security (which can become partly taxable depending on total income).

Instead of trying to perfectly predict the tax code for the next 30 years, add a planning placeholder: estimate taxes as a percentage of your retirement spending need. For many middle-income households, a rough placeholder might be 10%–20% of planned spending, adjusted as you learn more about your mix of accounts.

Maria and Devin do this simply: they add 15% on top of their spending target. Their Go-Go spending of $107,200 becomes roughly $123,000 of gross cash flow need. That number still might not all come from investments, but it's honest. It acknowledges that the IRS is a roommate in retirement.

6.9 The ``one-time'' expenses that deserve their own bucket

Some retirement costs are not annual lifestyle spending. They're lumpy, and if you pretend they're monthly, your budget becomes either too high to trust or too low to be safe.

Common lumps:

- replacing a roof, HVAC, or car,
- helping with a grandchild's education or wedding,
- a major move (closing costs, repairs, furniture, relocation),
- a dream trip you don't plan to repeat,
- dental or vision work not fully covered.

The fix is elegant: create a separate "lumpy spending" fund in your plan. Maybe it's $30,000, maybe $150,000, depending on your life. But it's not the same as your annual burn rate. It's a buffer that keeps a normal year from becoming a crisis year.

Maria and Devin create a $90,000 lumpy bucket for the first decade of retirement: one vehicle replacement, one major home project, and one family/helping-kids event. Seeing it explicitly makes them less likely to sabotage their own plan later by pretending those things "won't happen."

6.10 Retirement planning gets easier when you name the life you want

If you want a retirement number that actually works, stop asking, *"How much money do I need?"* Start asking questions that force specificity:

- *Where will Tuesday happen?* Same house, different state, closer to family, or closer to airports?
- *What gets better when work ends?* Time with people, health routines, volunteering, a side business, learning.
- *What do you refuse to give up?* Travel, gifting, eating well, living alone, staying near friends.
- *What are you quietly hoping goes away?* A second car, a high-cost zip code, stress spending.

These aren't dreamy questions. They are budget questions in disguise.

And they turn retirement from a vague threat into a series of trade-offs you can actually choose.

6.11 Actionable takeaway: build a ``Retirement Budget in One Hour'' draft

Take one hour this week and produce a first draft that's good enough to steer by.

1. Pull the last 3 months of spending from your bank and credit card.
2. Annualize the obvious categories (multiply by 4) and don't overthink the noise.
3. Create three columns: Go-Go / Slow-Go / No-Go.
4. Change only five categories at first: housing, transportation, healthcare, travel/leisure, and savings (remove retirement contributions from "spending" if you included them).
5. Add a taxes placeholder (10%–20%).
6. Write one sentence at the top of the page describing the retirement you're pricing: *"Stay in our home, travel twice a year for a decade, help family when needed."*

When your retirement number is built from your real life, it stops being a verdict. It becomes a plan you can negotiate with—and that's exactly what you want when time suddenly feels expensive.

6.12 The Healthcare Planning Gap

People don't usually retire because a spreadsheet told them to. They retire because something real happens: a layoff dressed up as a "reorg," a parent who suddenly needs help on Tuesdays, knees that don't love ladders anymore, a grandchild who makes time feel expensive. Then the calendar speaks up: *Medicare starts at 65.* If you're thinking about stepping away at 60, 62, even 63, that five-year stretch isn't just "early retirement." It's a coverage canyon.

And it's the canyon that quietly breaks otherwise solid retirement plans.

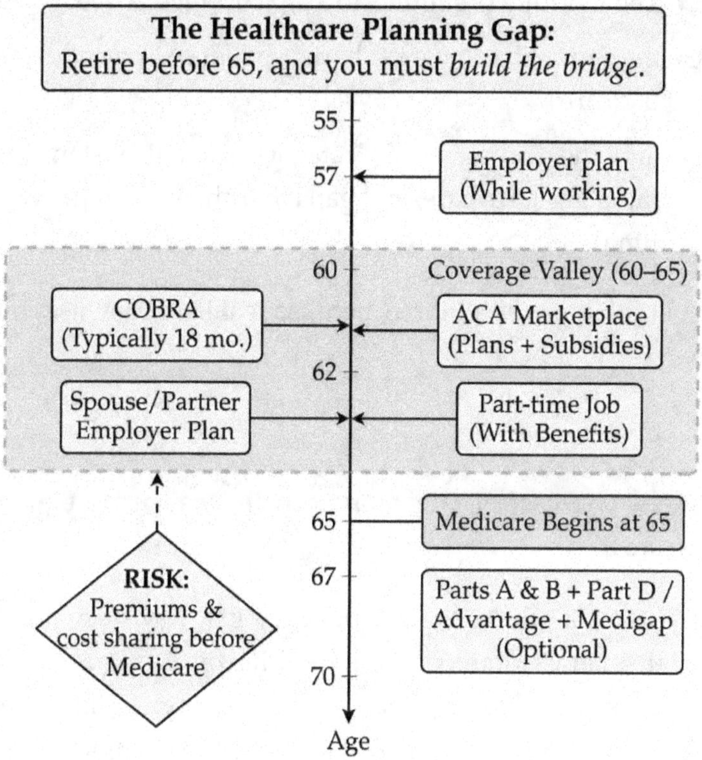

The reason this gap is so dangerous isn't that people forget health insurance exists. It's that they mis-price it. While you're employed, your premium is often hidden in plain sight: deducted pre-tax, subsidized by an employer, negotiated through a group plan that you didn't have to shop for. It feels like a utility bill. Then you leave work and discover the real sticker price—and the even more important detail: the premium is only the cover charge. The main expense can be what you pay *after* you're insured.

That's the mental shift that matters: healthcare in retirement is a three-part cost.

- The monthly premium (to keep coverage active),
- the deductible and cost sharing (what you pay when you use care),
- and the "not covered" category (dental, vision, hearing, long-term care, and gaps that show up at the worst time).

A solid plan prices all three, because real life rarely uses only one.

A real-world case: The early retirement that became a part-time job

Daryl is 62, a sales manager, and tired in a way sleep doesn't fix. His plan is common: work until his birthday, take Social Security at 62 to cover basics, and use savings to fill the rest until Medicare kicks in at 65. He and his spouse have about $780,000 saved, a paid-off house, and what he describes as a "reasonable" lifestyle.

Then he asks HR about continuing health coverage.

COBRA is available, but the premium is the full, unsubsidized cost of the employer plan plus a small administrative fee. The number is more than his mortgage used to be. If he keeps the same coverage for 18 months, it eats a chunk of the bridge money he thought was for everything else. If he switches to an ACA marketplace plan, the premium can be lower—but only if his taxable income is managed carefully enough to qualify for subsidies. If he takes Social Security, draws from a traditional 401(k), and does a few consulting projects "for fun," his income could jump just enough to make the ACA plan dramatically more expensive.

The twist is what changes his plan: he doesn't go back full-time. He takes a part-time role with benefits. Not because he loves the job, but because the health coverage is worth thousands per year, and it stabilizes the risk. His "early retirement" becomes "semi-retirement" largely because of one line item.

This is not a failure of planning. It's planning colliding with the most volatile expense in the household budget.

The bridge options: how people actually cover ages 60--65

Most people build the bridge one of five ways. The point isn't to crown a winner; it's to know which levers you can pull.

COBRA: Simple, familiar, and often expensive. It can be a good short-term move if you're in the middle of treatment, you've already hit deductibles, or you value continuity. The hidden risk is treating it like a multi-year solution. For many, it's a runway, not a road.

ACA marketplace plans: Potentially powerful, especially if you can qualify for premium tax credits (subsidies). The ACA turns healthcare planning into tax planning, because subsidies are based on your income. That means you may have real control, but you must be intentional about where your spending comes from.

Spouse or partner's employer plan: The cleanest bridge when it's available. It's also a reminder that retirement decisions are rarely individual. Sometimes one spouse works longer primarily to carry the health insurance.

Part-time work with benefits: Not glamorous, surprisingly effective. A job doesn't need to fund your whole lifestyle to be valuable; it can simply fund the healthcare line.

Private individual plans outside the ACA: Sometimes available, often less attractive, and can come with limitations. It's usually a last resort unless you're in a niche situation.

No matter which bridge you pick, treat it as its own mini-project. Healthcare isn't just another bill; it's a contract with rules, networks, and timing.

Medicare at 65: not ``free," not simple, but knowable

Medicare is often talked about like a finish line: once you hit 65, healthcare becomes easy. What really happens is that the pricing model changes from employer-based coverage to a menu of parts and add-ons.

A few numbers make Medicare feel real instead of mythical. In 2026, the standard Medicare Part B premium is $202.90 per month, and the Part B annual deductible is $283. Medicare Part A is premium-free for most people, but Part A cost sharing can still be substantial: for example, the inpatient hospital deductible is $1,736 per benefit period in 2026. Medicare Part A and Part B costs are published publicly and change over time, which is exactly why you want them in your plan as *adjustable inputs*, not vague assumptions.

Then there's the detail that surprises people who assumed Medicare works like employer insurance: *Original Medicare doesn't have a yearly out-of-pocket limit unless you have supplemental coverage or enroll in a Medicare Advantage plan.* That doesn't mean your spending will be infinite. It means the worst-case scenario can be much worse than people instinctively budget for, especially in a year with hospitalization, outpatient procedures, or expensive specialists.

This is where retirees typically choose between two broad paths:

- *Original Medicare + (optional) Medigap + Part D drug coverage,* often valued for broad provider access and predictable cost sharing if Medigap is robust.
- *Medicare Advantage (Part C),* which packages coverage and includes an out-of-pocket maximum for Part A and B services, but often uses networks and prior authorization.

The best choice is less about ideology and more about your tolerance for networks, your health needs, and how much you value cost predictability.

IRMAA: when your income makes Medicare more expensive

Here's the part that makes high earners blink: Medicare premiums can rise with income. The income-related monthly adjustment amount (IRMAA) is a surcharge added to Part B (and Part D) premiums for people above certain income thresholds.

This matters for late savers because the very moves that help you catch up—selling a highly appreciated asset, taking larger IRA distributions, doing big Roth conversions—can also push income up in a way that triggers higher Medicare premiums later. It's not a reason to avoid smart tax moves. It's a reason to *sequence* them.

If you're going to do income-spiking strategies, doing them before Medicare begins can be cleaner. If you're already on Medicare, you can still plan, but you plan with more speed bumps.

The HSA: the most misunderstood retirement account

If you have access to a Health Savings Account (HSA) through a high-deductible health plan while you're working, it may be the closest thing in personal finance to a legal cheat code.

Why? Because an HSA can be *triple tax-advantaged*:

- Contributions can be tax-deductible (or pre-tax through payroll),
- the money can grow tax-free if invested,
- and qualified medical withdrawals are tax-free.

The mistake many people make is treating the HSA like a checking account: they swipe it for today's co-pays because it's there. That's fine, but it misses the strategic opportunity.

A more powerful approach (when cash flow allows) is to pay current medical costs out of pocket and let the HSA grow. Save the receipts. Those receipts can often be used later to reimburse yourself tax-free, effectively turning old medical expenses into a future source of cash flow. Meanwhile, in retirement, the HSA can help cover qualified healthcare expenses that would otherwise force taxable withdrawals from IRAs.

This is what "healthcare planning" looks like when it's done well: you're not only buying insurance, you're building *flexible, tax-smart capacity* to handle whatever your body does next.

The costs people forget: dental, vision, hearing, and ``small'' big bills

Healthcare budgets get wrecked less by the obvious monthly premium and more by the expenses that feel too boring to model.

Dental is the classic example. Many employer plans include at least basic dental coverage, and many retirees discover the gap when they need crowns, implants, or periodontal work. Vision and hearing are similar: the need is common, the costs can be meaningful, and people tend not to include them because they don't happen monthly.

Then there's long-term care—which is often discussed as a separate category, but it belongs in the healthcare risk conversation because it can turn a family's best retirement plan into a caregiving plan. Even if you don't buy long-term care insurance, you can plan for the risk by deciding *where* the money would come from if home care or assisted living becomes necessary. Not because you expect it, but because you don't want to make financial decisions while making emotional ones.

A practical way to price the gap: build a ``bridge budget''

The most useful mindset for ages 60–65 is to treat healthcare like a bridge project with bids. You don't build a bridge by guessing. You get numbers.

Create a one-page bridge budget with these rows:

- Premiums (COBRA or ACA),
- deductibles and expected out-of-pocket costs,

- prescriptions,
- dental/vision/hearing,
- a "bad year" reserve.

Then price it in two scenarios:

- *Routine year:* physicals, a few visits, typical prescriptions.
- *Rough year:* hospitalization, outpatient procedures, imaging, specialist care.

This isn't pessimism. It's realism. People don't go broke in retirement because they planned for routine years. They go broke because they planned as if rough years were impossible.

Actionable takeaway: three moves to close the healthcare planning gap

If retirement is within ten years, do these three things while you still have the leverage of time and employment:

1. *Run the ages-60-to-65 quote test.* Price your bridge options now, even if you're 55. Get real premium estimates for COBRA (ask HR for the full cost) and an ACA plan (use your state marketplace). The goal isn't to pick today—it's to stop guessing.
2. *Start treating your income like a dial.* If an ACA plan might be part of your bridge, build a plan for where spending will come from (cash, Roth, taxable, traditional IRA). The same spending, drawn from different accounts, can mean very different subsidy outcomes.

3. *Protect the HSA if you have one.* If you can afford it, stop using the HSA as a swipe card. Invest it, save receipts, and let it become the account that quietly pays for future healthcare without forcing extra taxable withdrawals.

When people say, "I want to retire early," what they usually mean is, "I want my time back." The healthcare bridge is how you buy that time without turning your savings into a slow leak.

6.13 Social Security Optimization

Social Security has an image problem. People talk about it like a refund check: *"I paid in, now I want mine."* That mindset makes claiming at 62 feel like victory—a cash-out, a payback, a little revenge on every FICA deduction you ever saw.

But Social Security isn't a refund. It's an income stream with a built-in raise schedule, inflation adjustments, survivor protections, and rules that reward patience in a way the private market rarely does without charging you for it. Treated correctly, it functions like *longevity insurance*: the longer you live, the more valuable it becomes that the check keeps arriving, month after month, even if markets wobble or your portfolio has a bad decade.

That's why the claiming decision is so high-impact for people who feel late to retirement. You may not have thirty years left to compound contributions, but you can still choose how large your guaranteed baseline income will be for as long as you're alive.

The core trade: smaller checks sooner vs. bigger checks forever

Here's the simple truth hiding under all the rules: claiming earlier locks in a permanently smaller monthly benefit. Claiming later locks in a permanently larger one.

For people born in 1943 or later, Social Security adds *delayed retirement credits* of 8% per year (credited monthly) for waiting beyond full retirement age, and those credits stop accumulating at age 70. That creates a clean, powerful incentive: if you can afford to wait, you buy yourself a larger check for the rest of your life. If you can't afford to wait, you take the smaller check and solve for the gap elsewhere.

The most important reframing is psychological: you're not choosing the age that feels good. You're choosing the *size of your floor*.

Why the ``76% bigger" idea is both true and misleading

You'll often hear some version of: *"If you claim at 62 instead of 70, you're giving up about a 76% increase."* That's not just marketing; it's a reasonable ballpark comparison in many real cases. The ingredients are straightforward:

- Claim at 62 and your benefit is reduced relative to full retirement age (FRA).
- Delay past FRA and you earn credits (up to age 70).

But the phrase can mislead in two ways.

First, the actual percentages depend on your birth year and your FRA (which is 67 for people born in 1960 or later).

Second, what matters to your retirement plan is not the percentage increase—it's what the check does to your life. A $600/month difference is annoying. A $1,200/month difference changes whether you can keep your house, how much you can withdraw from investments, and whether your spouse can afford to be widowed.

That's why optimization is less about winning an argument on the internet and more about building the most resilient cash-flow design you can.

Break-even analysis: the picture that calms (or clarifies) the debate

When people argue about Social Security claiming, they usually argue about fairness: *"But if I wait, I'm missing years of checks!"* True. Waiting means you collect fewer total payments. The only question is whether the bigger payments later make up for the missing early years, and when.

That "when" is the break-even age—the point where total dollars received from the later-claiming strategy catches up to the early-claiming strategy. If you live past that point, delaying was financially rewarding in a simple dollars-and-cents sense. If you don't, claiming earlier may have paid more.

In real life, it's not that clean because taxes, investment returns, and inflation exist. But as a first pass, the break-even view is the best antidote to emotional claiming.

A simple model (clean on purpose)

Assume a worker whose full retirement age benefit is $2,000 per month at age 67 (FRA). Then:

- Claiming at 62 might be roughly $1,400/month (a reduction of about 30%).
- Claiming at 70 might be roughly $2,480/month (about 24% higher than FRA, reflecting 8% per year for three years).

Those are not personalized estimates. They're deliberately round numbers to show the shape of the trade.

Cumulative payout chart

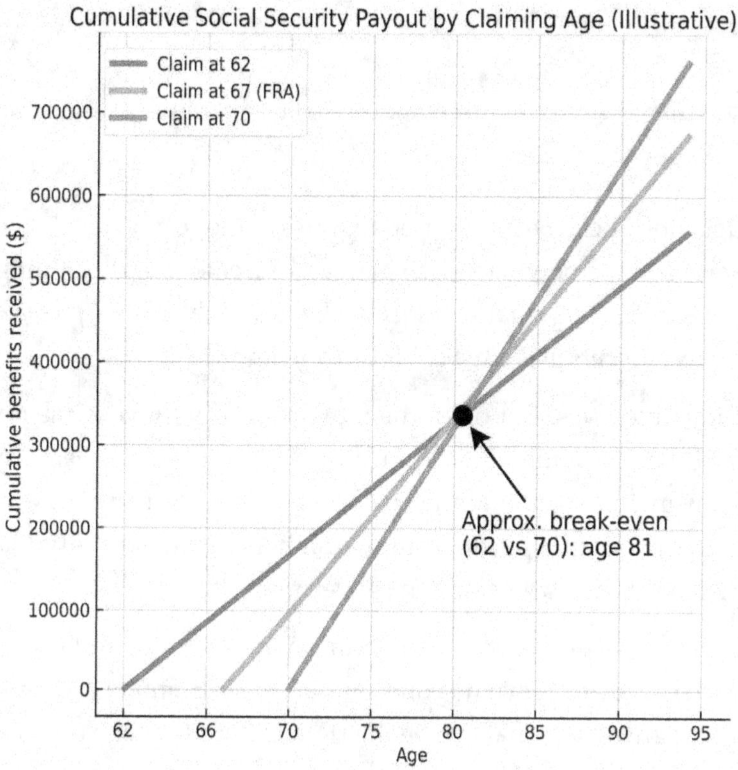

In this clean model, the lines cross in the early 80s. That's the entire argument in one image: if you think you're likely to live into your 80s (or you need to protect a spouse who might), waiting becomes less like "missing checks" and more like buying a bigger pension.

Now add the real-world kicker: Social Security benefits generally receive cost-of-living adjustments. Starting with a higher monthly benefit often means getting larger inflation adjustments in dollar terms, too. That's one reason high earners and higher savers frequently benefit from delaying when they can afford it.

The planning move that makes delaying possible: spend from your portfolio first

Delaying Social Security often requires a bridge: you need cash flow from somewhere else between retirement and claiming. For late starters, that can feel backward. *"Shouldn't I preserve my investments and take Social Security instead?"*

Sometimes, yes. But often the more powerful move is the opposite: use savings first to buy time, and let Social Security grow into a larger guaranteed check. In effect, you are converting part of your portfolio into a higher lifetime annuity-like income stream backed by the government.

This is not a universal rule. If your savings are thin and taking Social Security prevents you from raiding retirement accounts too aggressively, that can be the right call. But when you have the capacity, *spend down a bit early to lock in a bigger check later* can make your 80s and 90s dramatically safer.

A case study: Nora and James, and the survivor-benefit blind spot

Nora (63) earned significantly more than James (61) for most of their marriage. James took time out of the workforce when their kids were young and later worked part-time. Their retirement savings are decent but not lavish, and Nora is eager to claim at 62 because she doesn't like the idea of leaving money on the table.

Then they run the numbers in the way couples rarely do: *what happens when one of them dies?*

Here's the uncomfortable truth: the household will usually lose one Social Security check at the first death, keeping only the larger of the two benefits (the survivor benefit). That means the higher earner's claiming decision can become the surviving spouse's financial backbone.

If Nora claims early and locks in a permanently reduced benefit, she may be choosing a smaller survivor benefit for James later. If Nora delays (even partially), she's not just increasing their future income; she's increasing the insurance policy that protects the person likely to live longer.

Once they see it that way, the decision changes. They decide that James will claim earlier (because his benefit is smaller and they need some cash flow), while Nora delays as long as they reasonably can. Nora's larger benefit becomes the "keep-the-lights-on" check if she dies first. That's optimization that matters: not winning the most money in a vacuum, but preventing a surviving spouse from being forced into a move they didn't choose.

Working while claiming: the earnings test that confuses almost everyone

Another reason people claim early is that they plan to keep working "a little." That can be fine—but it comes with rules that are widely misunderstood.

If you claim Social Security before reaching full retirement age and you continue to earn wages, Social Security may withhold some benefits under the retirement earnings test. The key word is *withhold*, not *tax*. People hear "penalty" and assume the money is gone forever. It's more nuanced.

For 2026:

- If you are under full retirement age for the entire year, Social Security withholds $1 in benefits for every $2 you earn above $24,480.
- In the year you reach full retirement age, Social Security withholds $1 for every $3 you earn above $65,160 (counting only earnings before the month you reach FRA).

And there's a rule that's almost tailor-made for semi-retirement: the *special monthly earnings rule*. If you claim mid-year and have a few low-earning months, you can still receive benefits in those months even if your annual earnings exceed the annual limit. In 2026, the monthly thresholds are $2,040 for people under FRA all year, and $5,430 for those reaching FRA in 2026 (with additional conditions for self-employment).

This matters because it turns an all-or-nothing fear into a planning tool. You can stop full-time work in June, claim benefits, and still receive checks for months where you truly are retired (or close to it), even if the year's total earnings look high because of income earned earlier.

Also important: once you reach full retirement age, the earnings test goes away. Earn what you want; benefits are not reduced because of wages.

One more lever people ignore: continued work can raise your benefit

Social Security is based on your earnings record, not your last job title. If you keep working and earning, those new higher-earning years can replace earlier lower-earning years in the benefit calculation, increasing your eventual monthly benefit. For someone who had lean years in their 20s or 30s, or time out of the workforce, this can be a quiet advantage of working longer.

There's also a ceiling to keep in mind: Social Security taxes apply only up to the taxable maximum (the earnings cap). In 2026, that maximum taxable earnings amount is $184,500. Earnings above that level aren't subject to Social Security tax and generally don't increase the Social Security benefit formula in the same way.

The practical point isn't to memorize the cap; it's to understand that for many people, a few more strong earning years can improve the check you're optimizing in the first place.

Optimization isn't only about age: it's also about household strategy

A few household realities make the "best" claiming age different for different people:

Married couples: Often benefit from coordinating, not matching. A common pattern is lower earner claims earlier (to bring in some income), higher earner delays (to maximize the survivor benefit). It's not romantic, but it's protective.

Divorced individuals: If the marriage lasted long enough and other requirements are met, ex-spouse benefits may be available without affecting the ex-spouse's benefit. The claiming rules can differ from standard spousal timing, so it's worth checking the specifics before you assume you have no options.

Single retirees: The decision becomes more purely personal: life expectancy, cash-flow needs, and risk tolerance matter even more because there is no survivor-benefit planning to do. That can make delaying feel harder—and also make the "bigger guaranteed floor" even more valuable if you live long.

The theme is consistent: you're not merely choosing *when* to claim. You're choosing *how much guaranteed income you want in the years when you're least able to earn.*

Actionable takeaway: a two-meeting Social Security plan

Do this in two short sessions, not an endless research spiral.

1. *Meeting 1 (30 minutes): get your real numbers.* Create or log into your Social Security account and write down the estimated monthly benefit at three ages: 62, your FRA, and 70. Don't guess. Use the estimates tied to your earnings record.
2. *Meeting 2 (45 minutes): design a claiming plan that protects your future self.* For each age option, answer two questions:
    - What fills the cash-flow gap if you delay (which accounts, which amounts, for how long)?

- If you have a spouse, what does this choice do to the survivor benefit if the higher earner dies first?

Then pick a *default* claiming age and a *trigger* that would change it (job loss, health diagnosis, unexpected caregiving responsibilities). A default without a trigger is brittle; a default with a trigger is a plan.

The goal isn't to squeeze Social Security for every theoretical dollar. The goal is to make sure the check you'll depend on in your 80s is as strong as you can reasonably make it, because that's when financial flexibility is hardest to manufacture and peace of mind is worth the most.

6.14 Measuring the Gap: Scenario Planning

A retirement plan can look comforting right up until you ask one rude question: *Compared to what?* Compared to your actual spending. Compared to a bad market decade. Compared to the year you decide your knees are done with stairs. Compared to the day one of you dies and the household shifts from "we" to "me."

That's what scenario planning is: not pessimism, not spreadsheet theater—just the discipline of testing your plan against life's most common plot twists. You already did the hard, clarifying work: pricing the lifestyle you want and getting serious about healthcare and Social Security. Now comes the part where everything stops being abstract.

You measure the gap.

The gap is not your savings shortfall. It's your income shortfall.

Most people frame the problem like this: *"How much money do I need to have saved?"* It's a natural question, but it invites the wrong kind of answer—a single giant number that feels either impossible or meaningless.

A more useful question is: *"How much annual income does my plan require, and where does it come from?"*

Think of retirement income as a table with legs:

- *Guaranteed legs*: Social Security, pensions, certain annuities.
- *Flexible legs*: withdrawals from savings (401(k), IRA, brokerage, Roth), part-time work, rental income.

The gap is the space between the lifestyle you priced and the guaranteed legs you can count on. Your savings doesn't need to fund your entire retirement. It needs to fund the portion that isn't already covered by guaranteed income—and it needs to do it through good years and rough ones.

A clean worksheet that makes the gap visible in ten minutes

You don't need fancy software to find the gap. You need a simple sequence:

1. *Your annual spending target* (from your Go-Go/Slow-Go/No-Go budgets). Use the Go-Go number if you want to stress-test.
2. *Add taxes and healthcare reality checks* (not perfection; a credible placeholder).

3. *Subtract guaranteed income* (Social Security, pensions).
4. *The remainder is what your portfolio must cover.*

Then you convert that remainder into a portfolio target using a withdrawal rate assumption.

A withdrawal rate is just a planning dial: the percentage of your portfolio you expect to spend in the first year of retirement. The famous dial is 4%, but late starters often benefit from running more than one dial setting. Markets don't care what you expected.

A practical pairing:

- 4.0% as a baseline
- 3.7% (or 3.5%) as a more conservative stress-test

The point isn't to anoint a single "correct" withdrawal rate. It's to see whether your plan is robust or fragile.

Case study: Patrice and Eli, three scenarios, one honest gap

Patrice is 56 and Eli is 58. Patrice is a school administrator. Eli works in IT. They feel "behind," which is the most common financial emotion after 40: not broke, not wealthy, just uncomfortably aware that the runway is shorter than it used to be.

They've done the lifestyle pricing and land on these annual spending targets (in today's dollars):

- Go-Go (roughly 62–72): $110,000
- Slow-Go (roughly 73–80): $95,000

- No-Go (81+): $100,000 (less travel, more health and help-at-home)

They decide to test using $110,000 as the stress-test number, and they add a taxes placeholder of 15%. That pushes their gross cash-flow need to about $126,500.

Now guaranteed income. They look up their Social Security estimates and make a deliberate choice: they'll model Patrice claiming at 67 and Eli delaying to 70 to strengthen the survivor benefit. Their combined planned Social Security at those ages is estimated at $54,000 per year.

So the portfolio-funded gap at their "Go-Go + taxes" level is:

$$\$126{,}500 - \$54{,}000 = \$72{,}500 \text{ per year}$$

Now the portfolio target:

$$\text{At } 4\%\text{: } \$72{,}500/0.04 = \$1{,}812{,}500$$

$$\text{At } 3.7\%\text{: } \$72{,}500/0.037 \approx \$1{,}959{,}459$$

They currently have $920,000 invested for retirement.

That's the moment that would normally trigger either dread or denial—until you do the most important move in late-stage planning: you stop looking at it as one scenario. You build three.

Scenario 1: The Lean Retiree (spend less later)

This isn't "eat beans forever." It's identifying which parts of the Go-Go budget are truly optional and which are identity-level non-negotiables.

Patrice and Eli decide that $110,000 is their "nice" Go-Go year, but $100,000 is their "still happy" Go-Go year. That reduces the gross cash-flow need:

$$\$100{,}000 \times 1.15 = \$115{,}000$$

Then the gap becomes:

$$\$115{,}000 - \$54{,}000 = \$61{,}000$$

Portfolio target:

$$\$61{,}000/0.04 = \$1{,}525{,}000$$

Suddenly the gap between current assets ($920,000) and target ($1.525M) is not "impossible." It's a defined problem with levers.

The important emotional result is not the number. It's that they can see a lifestyle-based trade: spending $10k less per year reduces the required portfolio by hundreds of thousands. That's leverage you can feel.

Scenario 2: The Just-One-More-Year (retire later)

This is the late starter's secret weapon because it works in three directions at once:

- You save for one more year.
- Your portfolio has one more year to grow.
- Your retirement horizon shrinks by a year, so the plan has less time to fund.

Patrice and Eli run the math for retiring at 64 instead of 63, and then again at 65 instead of 63. The goal isn't to declare that everyone should work longer. The goal is to measure the impact, because impact is what changes behavior.

They discover that delaying retirement by two years makes their Social Security plan easier (less bridge spending) and lowers the amount they need to withdraw early from their portfolio. It also gives them more time to maneuver healthcare coverage before Medicare.

Two years doesn't just add dollars. It reduces fragility.

Scenario 3: The Aggressive Saver (save more now)

Aggressive saving after 40 is rarely about austerity. It's about removing the silent drains and redirecting the cash.

Patrice and Eli find $1,200 a month by making three targeted changes:

- One car becomes a "drive it into the ground" car instead of a "trade up every few years" car.
- A subscription audit that eliminates the small, stupid leaks.
- A planned dining budget that replaces the accidental dining budget.

$1,200 a month is $14,400 a year. Over 7–9 years, with investment growth, it becomes real money. Not magic money, but the kind that closes gaps.

The deeper value is psychological: they stop feeling like retirement is something happening *to* them and start feeling like it's something they're building.

The rule-driven cliffs that scenario planning catches

Scenario planning isn't only about markets. It's also about cliffs—thresholds where a small change creates a big effect.

A few common ones:

Social Security earnings test: If you claim before full retirement age and keep working, benefits may be withheld if earnings exceed certain limits. If you expect part-time income, your plan should model that. In the year you reach full retirement age, the rules change again. These are not moral judgments; they're mechanics. If you ignore them, the surprise arrives as a smaller deposit than you expected.

Medicare premiums and income: Higher income can mean higher Medicare premiums later due to income-related adjustments. A big Roth conversion, a large IRA withdrawal, or capital gains from selling an asset can create a ripple. Your plan doesn't need to forecast every detail, but it should recognize that income spikes can have side effects.

Original Medicare cost exposure: Without supplemental coverage, Original Medicare doesn't cap yearly out-of-pocket costs. That makes healthcare a natural stress-test variable. Your scenarios should include at least one "rough health year" assumption, because ignoring it is effectively betting your retirement on uninterrupted luck.

RMDs (required minimum distributions): Later in retirement, tax-deferred accounts force distributions. That can raise taxable income even if your spending doesn't rise. Scenario planning should test whether forced distributions could push you into higher taxes, affect Medicare costs, or change how much you want to keep in traditional accounts versus Roth.

These cliffs are exactly why a plan that looks fine at age 62 can feel chaotic at 75. Scenario planning keeps you from discovering the rules mid-flight.

Sequence-of-returns risk: the bad decade that matters most

If there's one market concept worth understanding before retirement, it's this: the average return is not the problem. The *order* of returns is.

If the market drops early in retirement while you're withdrawing, you sell more shares when prices are low. That can permanently reduce the portfolio's ability to recover. The same long-term average return can produce radically different outcomes depending on the first five to ten years.

Scenario planning deals with this by testing a "bad early sequence":

- Start retirement with a market decline (or flat returns) for 2–3 years.
- Keep withdrawals going (because life doesn't pause).
- See if the plan still survives.

If that scenario breaks your plan, you don't need to become a market prophet. You need a withdrawal strategy with flexibility: spending guardrails, a cash buffer, or a willingness to earn some income in early retirement if markets are ugly.

The goal is not to predict the next crash. The goal is to be the kind of retiree who doesn't get forced into bad decisions by one.

Three levers that actually close the gap (and one that only feels good)

When people see a gap, they often reach for the lever that feels heroic: picking more aggressive investments. That lever can help, but it's the least controllable. Markets don't sign contracts.

The controllable levers are almost always the same:

- *Save more now* (increase contributions, reduce leaks, capture employer match, use catch-up contributions if eligible).
- *Spend less later* (not deprivation—smart trade-offs in the categories that drive your number).
- *Work longer* (even part-time, even one or two years, because it hits the plan from multiple angles).

Here's the uncomfortable but liberating truth: for many late starters, working 12–24 months longer is more powerful than trying to out-invest the market. It's not because investing doesn't matter. It's because retirement is a cash-flow equation with timing, and timing is where work has disproportionate force.

A practical template: write your plan in plain English

A plan that only exists in a spreadsheet is fragile because you can't carry it around in your head. The best scenario planners can summarize their plan in a few sentences:

> *"We can retire at 64 if we keep spending near $100k, delay the higher Social Security benefit to 70, and keep a one-year cash buffer. If markets drop early, we pause travel spending and do part-time work for a year. If healthcare costs spike before 65, we use ACA coverage and limit income to protect subsidies."*

That is a real plan. It has triggers. It has flexibility. It acknowledges the rules without being ruled by them.

Actionable takeaway: build your three-scenario gap plan in one weekend

Set aside a weekend morning and do this with a pencil, not just an app:

1. *Write your spending targets* for Go-Go, Slow-Go, No-Go. Pick one as your stress-test number.
2. *Add two modifiers*: a taxes placeholder and a healthcare "rough year" placeholder.
3. *Subtract guaranteed income* using at least two Social Security claiming ages (an early option and a delayed option).
4. *Convert the remaining gap to a portfolio target* at 4.0% and 3.7% withdrawal rates.

5. *Create three scenarios* and write down exactly what changes in each: Lean Retiree: what spending drops, by how much; Just-One-More-Year: what age shifts, what that does to the gap; Aggressive Saver: what monthly amount increases now, and where it comes from.
6. *Choose triggers*: market drop, job loss, health event, family caregiving. Write the action you'd take in each.

When you're done, you won't just know your retirement number. You'll know the distance between where you are and where you want to be, and you'll have a few specific, realistic ways to close it—the kind that still work when life refuses to follow the neatest line on the chart.

Chapter 7

Converting Savings to Income

You've spent decades climbing the mountain of wealth accumulation, sweating over savings rates and compound interest. Now comes the trickier part: skiing down safely. Shifting from saving a paycheck to paying yourself from your portfolio is a psychological and mathematical pivot that trips up even savvy investors. This chapter isn't just about math; it's about engineering a paycheck that never bounces, protecting your nest egg when the market throws a tantrum, and making sure Uncle Sam doesn't take a bigger slice than necessary.

7.1 Surviving the ``Red Zone": Sequence of Returns Risk

Retirement has a marketing problem. For most of your working life, "the market" is a scoreboard. You check it, you wince, you cheer, you move on. But the day you start drawing income, the market stops being a scoreboard and becomes the conveyor belt that delivers dinner—or doesn't.

That change is why two retirees can do *everything* "right," invest in the same funds, earn the same average return over time, and still end up with wildly different outcomes. Not because one was smarter. Not because one was braver. Because one had bad timing.

The industry has a name for this: *sequence of returns risk*. Translation: *when* the returns happen matters as much as *what* the returns are. And the danger is concentrated in a short window that deserves a vivid name, because it behaves like one: the *Red Zone*.

The Red Zone is the five years before and the five years after retirement. Ten years where your financial life is unusually fragile. Ten years where a market drop isn't just a paper loss—it can become a permanent haircut. Ten years where you're most likely to make decisions under pressure: walking away from a job, selling a home, helping an adult kid, dealing with a health scare, or simply trying to figure out what "a normal month" costs when your paycheck disappears.

Why early losses are so much worse than late losses

Picture your portfolio like a fruit tree. When you're working, you're mostly planting: adding deposits, watering, letting it grow. In retirement, you're harvesting: pulling fruit every month to pay the bills. If a storm comes while the tree is young and you're still planting, you can keep watering and let time do its job. If a storm comes when you're harvesting heavily, you can strip the branches at the worst possible moment.

Here's the math underneath that metaphor, without the pain.

When markets fall, your investments are worth less. If you *don't* have to sell, you can wait for recovery. But retirement forces selling. Groceries don't accept "I'll pay you after the rebound." Your property tax bill doesn't care that the S&P is having feelings.

So a downturn early in retirement creates a double hit:

- Your portfolio drops.
- You still withdraw to live, which means selling more shares when prices are low.

Those shares are now gone forever. When markets recover, you participate with fewer shares. The recovery you were counting on has less engine to work with.

Later in retirement, the same downturn can be uncomfortable but less fatal. If you've had years of growth, your portfolio might be larger than where it started. You may also be drawing less from investments because Social Security has begun, a mortgage is paid off, or spending naturally slows. In other words, you have more cushion and often less pressure.

Alice and Bob: same average return, opposite fate

Alice and Bob retire at 65 with the same setup:

- Each has $1,000,000 invested in a balanced portfolio.
- Each withdraws $50,000 at the end of every year to fund spending (ignoring inflation for simplicity).
- Each experiences the exact same set of annual market returns over the next 10 years.

The only difference: the order of returns is reversed.

Alice gets the bad years first: her first two years are ugly (think of a bear market), and then she gets a recovery. Bob gets the recovery first, and then the bad years arrive later.

Their average return over 10 years is identical. The sequence isn't.

It feels like this *shouldn't* matter. It feels like life should average out. But withdrawals make it matter—a lot.

Here's a concrete (and intentionally simplified) set of returns to make the point: −20%, −10%, +15%, +15%, +10%, +8%, +6%, +6
Bob gets the same numbers in reverse order.

Alice's early losses force her to sell shares into the drop. By the time the good years arrive, her portfolio is smaller and has less power to rebound. Bob rides the early gains, so his later withdrawals come from a healthier base; when the bad years finally show up, he's bruised, not broken.

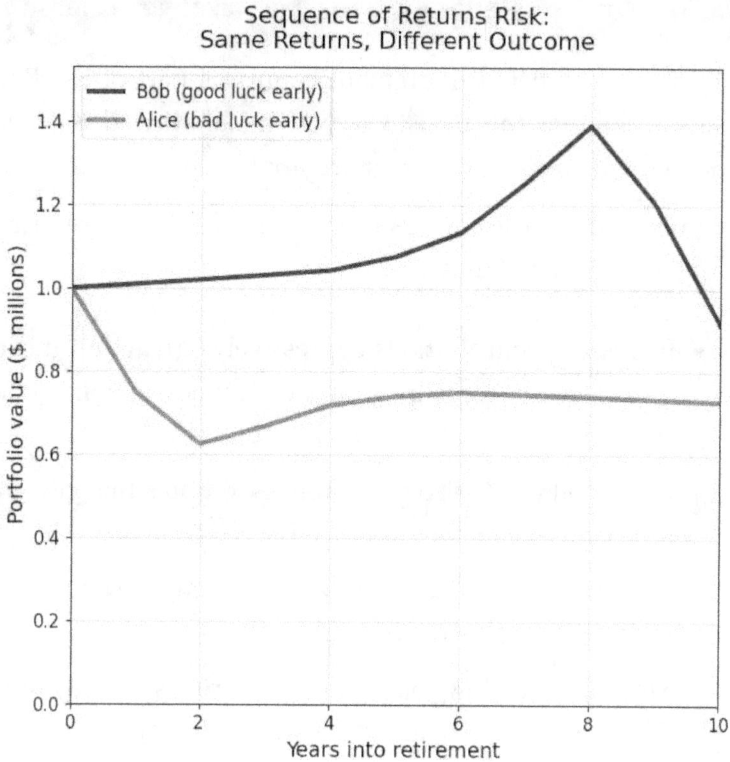

If that chart makes you slightly angry, good. It means you understand the unfair part: *Alice didn't do anything wrong.* She didn't "panic sell." She didn't buy meme stocks. She just retired into a storm.

That's why this risk deserves respect. It's not a test of virtue. It's a mechanical problem that requires mechanical defenses.

The Red Zone: a decade where small mistakes get expensive

Why five years before retirement counts as part of the Red Zone is simple: the decisions you make then determine whether you enter retirement sturdy or brittle.

If you're within a few years of stopping work, you're often doing at least one of these:

- Increasing contributions aggressively (great, but it can also increase exposure if all new money is going into stocks).
- Paying off debt (also great, unless it drains the only liquid cushion you had).
- Making a "final" asset allocation change (sometimes wise, sometimes driven by fear).
- Mentally spending the portfolio before it's actually time (the most common and the least discussed).

In that window, the market can hand you a rude surprise. A 25% decline on a $300,000 portfolio at 45 is painful. A 25% decline on a $1.5 million portfolio at 64, one year before you planned to retire, can rewrite the entire calendar of your life.

Then comes the five years *after* retirement, where sequence risk has sharp teeth. You're taking withdrawals, adjusting to a new routine, and you haven't yet built the emotional calluses that make a long retirement work. Early retirement is when people are most likely to spend like they're on vacation (because they are, at first) and most likely to feel trapped when markets fall (because they are, at least temporarily).

A real-world case study: retiring in 2000 versus 2010

Two couples, same savings habits, similar incomes, same target lifestyle. The only major difference: when they retired.

Couple A retired around 2000. The next decade included the dot-com crash and then the 2008 financial crisis. Their first years were loaded with drawdowns. Even if markets eventually recovered, their withdrawals during those declines locked in damage. Many people in this cohort responded in one of three ways: cut spending hard, go back to work, or accept a permanently smaller life.

Couple B retired around 2010. The next decade featured a long bull market with relatively mild pullbacks. Their early withdrawals came out of gains. They built confidence. They often increased spending (sometimes responsibly, sometimes not) because the portfolio seemed to "handle it."

This comparison isn't about predicting the next decade. It's about understanding what you can control. You can't choose your retirement "weather." But you can decide whether you're bringing a raincoat or showing up in flip-flops.

Defense #1: Build a cash buffer that protects your dignity

A cash buffer is not an "investment." It's an anti-panic device.

The goal is simple: if stocks are down, you want *permission* to not sell them. A dedicated pool of safe, liquid money gives you that permission. It turns a market crash from "I must sell now" into "annoying, but manageable."

A practical target for many households is 12 to 24 months of core spending in something boring: high-yield savings, money market funds, short-term Treasuries, or a conservative short-term bond fund if you understand the tradeoffs. The right number depends on:

- How stable your non-portfolio income is (pension, Social Security timing, part-time work).
- How flexible your spending is (can you cut travel, delay a car purchase, pause gifting?).
- How aggressive your portfolio is (more stocks generally means bigger swings).

The psychological benefit matters as much as the math. When the market drops 30%, your brain does not calmly compute expected returns. Your brain tells you a story. Usually a scary one. A cash buffer interrupts that story with a simple fact: *you can pay the bills without selling at a discount.*

One warning: a cash buffer is not a license to ignore the portfolio. It's a bridge, not a bunker. Its job is to buy time for recovery and give you flexibility to choose *when* to sell, rather than being forced.

Defense #2: The ``bond tent'': calmer early years, more growth later

A bond tent is a deliberate shape to your asset allocation around retirement. The idea is to increase your allocation to high-quality bonds (and/or cash-like assets) as you approach retirement, then gradually reduce that safety allocation over the early retirement years.

It's called a "tent" because the bond allocation rises into retirement (the peak of the tent) and then slopes downward afterward.

Why would you ever reduce bonds as you age? Because sequence risk is front-loaded. The first decade is when withdrawals and volatility collide. If you can survive that window without selling depressed stocks, the rest of retirement often becomes more forgiving. In those later years, growth can matter more again, especially for people who might live into their 90s.

This approach also sidesteps a common trap: shifting to a very conservative portfolio at 65 and then staying there forever. That can feel safe, but it can quietly create a different risk: *longevity risk* (the risk of running out later because the portfolio doesn't grow enough to keep up with decades of inflation).

The bond tent tries to keep you from being forced into the worst trade in finance: selling your growth assets right after they've been punched in the mouth.

Defense #3: Spend from the safe bucket when stocks are down (and refill when they recover)

A surprisingly effective retirement "system" is a two- or three-bucket approach that's less about fancy optimization and more about behavior you can stick with.

One straightforward version:

- *Spending bucket*: 1–2 years of withdrawals in cash or cash-like instruments.
- *Stability bucket*: high-quality bonds meant to refill the spending bucket.
- *Growth bucket*: diversified stocks for long-term growth.

When markets are down sharply, you draw from the spending bucket instead of selling stocks at low prices. When markets are strong (or after a recovery), you sell some appreciated assets from the growth bucket to refill the spending bucket.

This doesn't require predicting the market. It requires a rule you can follow. The rule can be as simple as: "If stocks are down more than X% from their recent high, don't sell stocks for spending this year."

That rule does two things:

- It reduces the mechanical damage of selling low.
- It reduces the emotional damage of feeling like you are locking in loss.

People underestimate the emotional damage. A retirement plan can be mathematically sound and still fail because the human living inside it can't tolerate the stress. A bucket structure lowers stress, which lowers the odds of the one move that ruins everything: capitulating near the bottom.

Defense #4: Delay retirement by months, not years (if you can)

This is the least glamorous lever, but it can be the most powerful in the Red Zone: a small shift in timing.

Working even 6–12 additional months can help in three compounding ways:

- You add more savings (one more year of contributions).
- You reduce withdrawals (one less year of drawing).
- You potentially improve Social Security outcomes (if delaying claiming is part of the plan).

It's also an unusually "clean" fix because it doesn't require market luck. It requires life logistics, health, and willingness—which are real constraints, but at least they're yours.

This doesn't mean everyone should delay. It means that if a bear market hits right at the edge of your planned retirement, the first question isn't "Should I sell everything?" The first question is "Can I buy myself a little time?" Time is the antidote to forced selling.

Defense #5: Decide in advance what you'll cut first

When people say "I can cut spending if I have to," they often mean "I'll panic and cut everything." That's not a plan. That's shock.

A better approach is to pre-choose a small list of "first cuts" that don't destroy your quality of life. Not theoretical cuts—real ones with dollar amounts.

For example:

- Pause travel for a year ($6,000–$12,000).
- Drive the current car two more years (delay a $35,000 purchase).

- Reduce gifting temporarily ($2,000–$10,000).
- Swap dining out frequency ($200–$600/month).

The point isn't austerity. The point is to protect the portfolio during its most fragile years with small, reversible moves rather than big, permanent ones.

If you've ever watched someone try to diet by cutting everything they enjoy, you know how this ends. Retirement spending works the same way. The plan must be livable, or it won't be followed.

How to know if you're exposed: a quick self-audit

Sequence risk loves three ingredients: high withdrawals, high volatility, and low flexibility. You don't need a Monte Carlo simulation to get a useful read on your exposure. You need honest answers.

- If the market fell 25% next month, could you cover 12 months of spending without selling stocks?
- Are you planning to withdraw more than about 4%–5% of your portfolio in the first year (before Social Security, pensions, or part-time income)?
- Is your portfolio concentrated (single stock, heavy sector bets, too much employer stock)?
- Would a spending cut feel like "annoying" or like "life-altering"?

If those answers make you uncomfortable, good. Discomfort is diagnostic. It tells you where to reinforce.

A simple, practical Red Zone plan

You don't need a complicated machine. You need a sequence-risk seatbelt. A workable plan for many retirees looks like this:

1. Hold 12–24 months of core spending in a safe, liquid bucket.
2. Keep a meaningful allocation to high-quality bonds as shock absorbers during the early years.
3. Maintain diversified equities so the plan can survive 25–30+ years.
4. Use a written rule for what you sell (or don't sell) when markets are down.
5. Pre-select a few spending adjustments that you can live with if a bad sequence hits.

The power isn't in any single piece. It's in the way they work together. Cash buys time. Bonds dampen the fall. Rules prevent forced selling. Spending flexibility reduces strain. Diversified equities keep the plan from quietly dying of old age.

Actionable takeaway: build your ``no-forced-selling'' year

Pick one date in the next two weeks and do one piece of work that future-you will feel in their chest as relief.

Calculate your *core monthly spending*—the boring, non-negotiable stuff: housing, utilities, groceries, insurance, basic transportation, healthcare. Multiply by 12. That number is your *no-forced-selling year*.

Then answer one question: where, exactly, will that money live during the Red Zone?

If the answer is "in the stock market," you've found the crack. Fill it with something boring and liquid. The goal isn't to beat inflation with that bucket. The goal is to make sure a bad year in the market doesn't become a bad decade in your life.

7.2 Engineering Your Paycheck: Withdrawal Strategies That Flex

A steady paycheck is one of adulthood's best inventions. It quietly solves a thousand decisions. You don't have to negotiate with the grocery store. You don't have to time the gas station. You just live, and the money shows up.

Retirement flips that. Now *you* are the payroll department, and your "salary" comes from a machine that sometimes throws sparks. The hard part isn't knowing that markets go up and down. The hard part is designing a spending plan that doesn't force you to pretend they don't.

That's why the famous "4% rule" became so popular. It gave people a number and, more importantly, a feeling: *I can stop thinking about this.* Withdraw around 4% of your starting portfolio in year one, adjust that dollar amount for inflation every year, and historically you often made it through a 30-year retirement.

As a starting point, it's useful. As an autopilot, it's a trap.

The trap is emotional as much as mathematical. A rigid rule turns every market decline into a personal crisis: "Am I still allowed to take my inflation raise?" A flexible rule turns the same decline into a managed event: "This is one of those years where the plan asks for a small cut." When you're living on a portfolio, the second mindset isn't just calmer—it's often safer.

The hidden promise inside the 4% rule (and why it's easy to misread)

The 4% rule is often described like a guarantee. It isn't. It's a historical observation based on specific assumptions: a diversified portfolio, a long time horizon, U.S. market history, and a retiree willing to take the same inflation-adjusted paycheck even when the portfolio is shrinking.

That last part is the quiet, brutal promise: *you will keep paying yourself the same real income even in ugly markets.* In other words, you're agreeing to be least flexible at the exact moment flexibility matters most.

Plenty of people can't do that, and it doesn't make them irrational. It makes them human. If you're watching the portfolio fall and you keep taking inflation raises anyway, you need nerves of steel or a kind of emotional distance most people don't have—especially when headlines are screaming and friends are delaying retirement.

There's another misread: people hear "4%" and think it's a safe *range*. Like 4.5% is basically 4%. Like 5% is just 4% plus confidence. But withdrawal rates don't behave like that. They behave like a bridge weight limit. Exceed it a little and you might still be fine. Or you might find the weak point.

Flexible withdrawal strategies exist because retirees don't actually want a rule; they want a paycheck that survives real life.

A retirement paycheck shouldn't be a straight line

Here's a thought experiment. Imagine your employer said: "Your salary will be the same every year, but the company's revenue will swing wildly. Some years we'll make record profits. Some years we'll lose money. Your salary won't change."

You wouldn't call that stable. You'd call it fragile. Eventually, something breaks.

A portfolio is the same kind of business. If returns are high, you can afford a raise. If returns are low or negative, giving yourself a raise anyway might be the financial equivalent of running the factory harder while the power is flickering.

A flexible retirement paycheck accepts a simple truth: *spending is a lever*. Not a moral lever—a practical one. Pull it gently in the tough years and you often avoid pulling it brutally later.

The goal is not to make retirement feel like a permanent austerity program. The goal is to avoid the two outcomes that ruin the experience:

- Overspending early and quietly sabotaging the later decades.
- Underspending for years because you're terrified of running out, then realizing too late you saved joy you could have afforded.

Flexibility is how you thread that needle.

The ``raise" you skip can be worth more than the raise you take

Most people picture flexibility as a dramatic spending cut—canceling retirement, selling the house, moving in with family. That's not what the research and real-world practice tend to rely on.

It's often smaller and more surgical: skipping inflation increases in down years, trimming discretionary spending for a short stretch, or pausing big one-time purchases. Those modest adjustments can have outsized effects because they happen when the portfolio is most sensitive.

Think of it like steering on ice. A small correction early prevents a skid that forces a major correction later.

This is also why flexible strategies can feel psychologically fair. When markets are strong, you can give yourself a raise or a "bonus." When markets are weak, you accept a temporary "pay freeze" or a modest cut. That's how most people intuitively think income should work. It maps to normal life.

Meet the Guardrails approach: rules you can live with

A flexible strategy needs structure. Otherwise, "flexible" becomes "I'll guess every year," which becomes "I'll worry every year." That's not retirement; that's a subscription service for anxiety.

One of the most practical frameworks is a guardrails approach. The idea is simple enough to explain on a napkin:

- Start with a target spending amount.

- Set an upper rail and a lower rail around that target.
- If your spending rate drifts too high because markets fell, you tighten spending (a pay cut).
- If your spending rate drifts low because markets rose, you loosen spending (a bonus).

The guardrails are triggered by your *withdrawal rate relative to the current portfolio*, not the original portfolio. That's the key. It keeps your spending from becoming disconnected from reality.

A common version looks like this (conceptually): start around 4%–5%, then set a "too high" rail around 5%–6% and a "too low" rail around 3%–4%. The exact numbers depend on your age, portfolio mix, and how much guaranteed income you have. The spirit is what matters: you pre-commit to action *before* the situation becomes desperate.

The brilliance of guardrails is that they give you permission in both directions. Many retirees only plan for bad markets. Guardrails also address the quieter risk: being unnecessarily stingy when the portfolio is thriving.

A concrete example: Maria engineers a paycheck she can actually follow

Maria is 66, recently retired, and has $900,000 invested across a diversified portfolio. She expects Social Security to cover about half of her core expenses, but not the fun stuff: travel, grandkids, hobbies, and the occasional "we deserve it" dinner.

She wants $45,000 per year from her portfolio to start. That's 5% of $900,000. It's not crazy, but it's ambitious—especially early.

So instead of pretending she can take $45,000 plus inflation every year forever, Maria builds guardrails:

- *Target*: $45,000 from the portfolio this year.
- *Upper rail (pay cut trigger)*: if $45,000 would be more than 5.8% of the portfolio value at the start of the year, she reduces withdrawals by 10%.
- *Lower rail (bonus trigger)*: if $45,000 would be less than 4.2% of the portfolio value at the start of the year, she increases withdrawals by 10% (or takes a one-time "bonus" for travel).

Now a rough year doesn't force her into an identity crisis. It forces her into a pre-agreed response.

Suppose the market drops and her portfolio begins the next year at $780,000. If she insisted on $45,000, that's 5.77%. She's brushing the upper rail. Under her plan, she reduces spending to $40,500. That $4,500 adjustment might be as simple as one less trip, delaying a kitchen remodel, or being less generous with gifts for a year.

Is that fun? No. Is it survivable? Yes. And survivable beats dramatic.

Now suppose a few strong years follow and her portfolio begins a later year at $1,050,000. Her $45,000 is now 4.29%. She's approaching the lower rail. Under her plan, she gets to increase spending—not because she feels euphoric, but because the rule says she can.

Maria doesn't need perfect forecasting. She needs a system that keeps her from making the two classic mistakes: pretending markets are smooth, and treating every market move like an emergency.

The point of guardrails isn't mathematical perfection---it's preventing regret

The best withdrawal strategy in the world is worthless if you abandon it mid-crisis.

Most retirees don't blow up their plans because they can't calculate. They blow up their plans because they can't sleep. A strategy that asks you to do something that feels insane at the worst possible moment—like taking an inflation raise while your portfolio is down 30%—may be mathematically coherent and behaviorally impossible.

Guardrails acknowledge that you are part of the equation. They build in a "pressure release valve." When markets are bad, you don't have to wonder what to do. You do the thing you agreed to do while you were calm.

That calm version of you is smarter than the stressed version. Guardrails are a way of letting calm-you stay in charge.

Other flexible approaches that work (and why they feel different)

Guardrails are not the only flexible strategy, but they're one of the easiest to operationalize. Two other common approaches are worth knowing because they match different personalities.

Spending floors and ceilings. Some people hate percentage triggers. They'd rather decide on a minimum acceptable lifestyle and a maximum "we're not being ridiculous" lifestyle. In practice, this means setting a floor ("no matter what, we can spend $X") supported by guaranteed income and safer assets, plus a ceiling that prevents lifestyle creep. It's less responsive than guardrails, but it can feel more humane.

Skip raises in bad years. This is the simplest rule that still captures much of the benefit of flexibility. You start with a baseline withdrawal and normally adjust for inflation, *except* in years when the portfolio is down. In down years, you hold spending flat (a pay freeze), and when markets recover you resume inflation adjustments. It's not as nuanced, but it's easy to explain to a spouse and easy to follow when stressed. That last part is not a small advantage.

The common thread: you're no longer demanding a straight-line paycheck from a jagged machine.

A caution: flexibility is not the same as permanent deprivation

Some people hear "flexible withdrawals" and imagine a retirement defined by constant belt-tightening. That's not what a well-designed flexible strategy produces.

A good flexible plan has three features:

- *Adjustments are occasional, not constant.* If you're changing spending every month, the rules are too twitchy.
- *Adjustments are meaningful but not life-shattering.* A 10% trim is uncomfortable. A 40% trim is a crisis.
- *Upside is real.* In strong markets, you should be allowed to live a little. Otherwise the strategy becomes emotionally one-sided, and people stop respecting it.

The goal is to smooth the consequences, not eliminate them. Markets still move. The plan just keeps those moves from dictating your entire lifestyle.

What makes a flexible strategy succeed: separating ``core" from ``nice-to-have"

Flexibility requires something most people avoid until a crisis forces it: naming your priorities.

If every dollar of spending feels equally important, then any cut feels like an attack on your life. But if you divide spending into *core* and *optional*, the plan becomes much easier to execute.

Core spending is the stuff that makes your life function: housing, utilities, food, insurance, healthcare, basic transportation. Optional spending is what makes life feel like *your* life: travel, gifts, hobbies, home upgrades, eating out, the expensive gym you genuinely love.

A flexible paycheck protects core spending like it's sacred. Flexibility mostly happens in the optional category. That's not just financially smart; it's emotionally protective. It keeps a down market from feeling like a threat to your safety.

Many retirees do the opposite by accident: they treat optional spending as sacred ("we promised ourselves travel") and then squeeze core spending with anxiety-driven decisions (skipping healthcare, delaying repairs, underheating the house). A guardrails mindset steers you away from that.

A real-world stress test: the first bad year

The first time your plan meets a genuinely bad market year is where theory becomes character.

If you've built flexibility in advance, the conversation sounds like: "Okay, this is one of those years. We're doing the 10% trim. What do we cut first?"

If you haven't, the conversation sounds like: "Do we sell? Do we stop traveling forever? Are we going to run out?" That second conversation has a way of producing expensive decisions: panic selling, abandoning diversification, or swearing off spending even when the portfolio later recovers.

One practical technique that helps: write down your planned adjustments *before* the storm. Not in a spreadsheet you'll never open, but in plain language.

For example:

> If the portfolio is down and we hit the upper rail, we will (1) pause large gifts, (2) take one local trip instead of flying, and (3) delay the car purchase.

That list is not a restriction. It's a relief. It means you've already decided what "tightening the belt" means, so you don't have to negotiate it while stressed.

Actionable takeaway: build a raise-and-freeze policy for your retirement paycheck

Choose one of these two rules and write it down—literally, on paper or in a shared note with your spouse/partner:

- *Simple version (raise-and-freeze):* "We take an inflation increase only in years when the portfolio is up. If the portfolio is down, our paycheck stays flat next year."
- *Structured version (guardrails-light):* "We target $X this year. If that amount would exceed Y% of our current portfolio value next year, we reduce spending by 10%. If it would fall below Z%, we increase spending by 10% or take a one-time bonus."

Then pick the one discretionary category you'll adjust first—the one cut that saves meaningful dollars with the least misery. When the market eventually hands you an ugly year (because it will), you won't be inventing a plan in the middle of the fire. You'll be following one.

7.3 The Tax Torpedo: Sequencing Your Withdrawals

People worry about running out of money in retirement. Reasonable fear. But there's another kind of "running out" that happens in slow motion: you keep the same lifestyle, your portfolio is fine, and yet your tax bill quietly inflates like a pufferfish. Not because Congress changed your bracket. Because your own withdrawals tripped a wire you didn't know existed.

That wire is what planners sometimes call the *tax torpedo*: the moment additional income doesn't just get taxed itself—it causes *other* income (especially Social Security) to become more taxable, and can also shove you over Medicare premium cliffs. The result is a surprising effective tax rate, right when you were hoping retirement would be simpler.

The frustrating part is that the torpedo isn't triggered by being reckless. It's often triggered by being orderly. People follow the common advice—"spend taxable first, then IRA/401(k), then Roth"—and later discover they've built a giant pre-tax account that will eventually force withdrawals at the worst possible time.

Withdrawal sequencing isn't about choosing the *right* account in a vacuum. It's about engineering your *lifetime* tax shape: keeping income steady enough to avoid spikes, low enough to avoid cliffs, and predictable enough that you're not surprised every April.

7.4 Three flavors of money, three kinds of tax consequences

Most retiree paychecks are blended from three account types:

- *Taxable (brokerage, savings, CDs)*: you already paid tax on the principal; taxes show up as interest, dividends, and capital gains.

- *Tax-deferred* (*Traditional IRA, 401(k), 403(b)*): you got a tax break going in; withdrawals are generally taxed as ordinary income.
- *Tax-free* (*Roth IRA, Roth 401(k)*): you paid tax going in; qualified withdrawals can be tax-free, and Roth IRAs have no required minimum distributions during the owner's lifetime (under current rules).

The mistake is treating these like three fuel tanks and asking, "Which tank do I drain first?" A better question is, "How do I keep my taxable income from lurching around as I age?"

Because taxes aren't just a rate table. They're a series of trip-wires.

7.5 Tripwire #1: Social Security taxation (it's not all-or-nothing)

Social Security feels tax-free when you first start it. Then you add a little IRA withdrawal or a little interest income and, suddenly, your benefits show up on the taxable line.

The IRS uses a measure commonly described as *provisional income*: roughly your other income plus *half* of your Social Security benefits, with tax-exempt interest included. IRS Publication 915 lays out the mechanics and the thresholds. (irs.gov)

The thresholds are famously *not* indexed for inflation, which is why more retirees get snagged over time. For many filing statuses, provisional income below \$25,000 (single) or \$32,000 (married filing jointly) generally means no federal tax on benefits; between those thresholds and \$34,000/\$44,000 can mean up to 50% of benefits taxable; above $34,000/$44,000 can mean up to 85% taxable. (irs.gov)

That "up to 85%" line gets misread. It doesn't mean you pay an 85% tax rate. It means up to 85% of your benefit can be included as taxable income. The torpedo happens because a $1 increase in IRA withdrawals can cause *more* than $1 to be taxed (because it drags some Social Security into the taxable column too). IRS Publication 915 explicitly describes the two-tier system and how additional income increases the taxable portion. (irs.gov)

Why it matters for sequencing: if you wait to touch your tax-deferred accounts until after Social Security starts, you can accidentally create years where every extra dollar from the IRA does double duty: it's taxed itself and it makes more of your Social Security taxable.

7.6 Tripwire #2: Medicare IRMAA (the cliff you don't feel until two years later)

IRMAA is Medicare's income-related surcharge: higher-income retirees pay more for Part B (doctor/outpatient coverage) and Part D (drug coverage). It's based on your modified adjusted gross income (MAGI) from *two years prior*.

That two-year lag is what makes IRMAA feel like a prank. You do a large Roth conversion at 66, feel virtuous, and then at 68 your Medicare premium jumps and you're staring at a letter that looks like a billing error.

CMS publishes the official IRMAA brackets. For 2026, the standard Part B premium is $202.90/month, and the Part B total premium climbs by tiers as income rises; the first IRMAA tier begins above $109,000 MAGI for single filers (above $218,000 for joint), with progressively higher premiums up to $689.90/month at the top tier. Part D has its own IRMAA add-on, from $14.50 to $91.00/month depending on tier. (cms.gov)

The key behavioral point is the cliff: a dollar over the threshold can trigger the full surcharge for that tier. CMS's tables make that structure plain. (cms.gov)

Why it matters for sequencing: if you plan withdrawals without watching MAGI, you can turn a one-time income spike (selling a property, big IRA withdrawal, large capital gain) into a recurring premium pain point for a full year.

7.7 Tripwire #3: Required minimum distributions (RMDs) turn ``maybe'' into ``must''

Tax-deferred accounts are patient, not forgiving. Eventually, the IRS requires you to start taking withdrawals whether you want them or not.

Under IRS guidance for current law, traditional IRA owners generally must begin RMDs at age 73 (for those subject to the age-73 rule), with a required beginning date of April 1 of the year after reaching the applicable age, and then RMDs due by December 31 in subsequent years. (irs.gov)

Two details from Publication 590-B matter in real life:

- The first-year timing can bunch income if you delay: taking the first RMD by April 1 means you may also need to take the second RMD by December 31 of the same calendar year. (irs.gov)

- Taking more than the RMD doesn't "bank" credit for later years. Extra withdrawals are just extra taxable income. (irs.gov)

Why it matters for sequencing: if you let your pre-tax accounts grow untouched for decades and then Social Security starts and then RMDs start, you can create a perfect storm: forced ordinary income on top of benefits on top of Medicare cliffs.

7.8 The comforting rule of thumb---and the reason it sometimes backfires

The classic advice goes like this:

> Spend taxable assets first, then tax-deferred, then Roth.

It's comforting because it sounds clean. It's also sometimes sensible. Taxable accounts can be tax-efficient (long-term capital gains rates, step-up in basis for heirs), and letting tax-deferred accounts compound can be valuable.

But here's the backfire scenario: by protecting the IRA for too long, you build a large future tax bomb. Later, RMDs force big ordinary-income withdrawals in the same years you're receiving Social Security. That combination pushes more benefits into taxable income and can trigger IRMAA surcharges. Publication 915 and CMS's IRMAA tables are the two pieces of "plumbing" that make this dynamic real. (irs.gov)

The goal isn't to declare the rule of thumb wrong. The goal is to stop treating it like a law of physics.

A more useful mental model is *bracket filling*: deliberately creating a steady, controlled amount of ordinary income each year, instead of starving yourself of income early and then being force-fed later.

7.9 A concrete case study: the couple who ``did everything right"

Consider Dan and Patrice, both 66, newly retired.

- $200,000 in a taxable brokerage account (mostly index funds with low cost basis).
- $1,300,000 in Dan's traditional 401(k)/IRA (rolled over).
- $150,000 in a Roth IRA.
- They plan to start Social Security at 70.

Their instinct is to live on taxable money for four years, then start Social Security, then touch the IRA "as needed."

Here's what happens if they follow that instinct without thinking about taxes:

1. Ages 66–69: taxable account withdrawals are partly principal (not taxed), so their taxable income looks low. They feel like they're in a "low bracket" and they enjoy it.
2. Age 70: Social Security begins. Now there's a new income stream that interacts with other income via provisional income rules. (irs.gov)

3. Age 73: RMDs begin. Now the IRA creates mandatory ordinary income. (irs.gov)
4. Mid-70s: the IRA is still large (it compounded for years), so RMDs climb. Those withdrawals can make up to 85% of Social Security taxable, and they may push MAGI into IRMAA tiers, raising Medicare premiums. (irs.gov)

Now watch the alternate approach. Dan and Patrice decide to use ages 66–69 as *intentional tax years*. They do partial Roth conversions each year: enough to "fill" a target tax bracket but not enough to trigger an IRMAA tier they want to avoid. They still use some taxable withdrawals for spending, but they refuse to let their taxable income drift too low, because they know low now can mean high later.

When Social Security starts at 70, their traditional IRA balance is smaller than it would have been. When RMDs begin at 73, the required withdrawals are smaller too. That can reduce the chance of pushing benefits into the higher taxable regime and reduce the odds of Medicare surcharges. (irs.gov)

They didn't avoid tax. They *scheduled* it.

That's the whole game.

7.10 A practical sequencing framework that respects the tripwires

There isn't one perfect order, but there is a practical way to make decisions that won't betray you later.

7.11 1) Cover the basics with the least ``tax collateral damage"

If you have spending needs and choices, prefer sources that don't cause a lot of ordinary income when you're trying to manage brackets:

- In taxable accounts, selling lots with higher cost basis can create cash with smaller capital gains.
- Cash and CDs don't create capital gains surprises, though interest is ordinary income.
- Roth withdrawals (if qualified) can be powerful because they may not increase taxable income—but that's exactly why you don't want to drain Roth thoughtlessly.

7.12 2) Use tax-deferred money on purpose (not by default)

Tax-deferred withdrawals are not "bad." They're just potent. Treat them like a spice: the right amount makes the meal; too much ruins it.

Many retirees benefit from taking *some* tax-deferred income earlier than they feel they need, especially in years before Social Security or before RMDs. Those years can be unusually flexible because you're not simultaneously managing benefit taxation and mandatory distributions. (irs.gov)

7.13 3) Treat the Roth IRA as your tax and IRMAA shock absorber

Roth money is uniquely useful later because it can provide cash flow without increasing taxable income (when withdrawals are qualified). That makes it a potential tool for:

- A year with a large one-time expense (roof, car, family support).
- A year where you're close to an IRMAA threshold and don't want to cross it. (cms.gov)
- A year where additional ordinary income would drag more Social Security into taxation. (irs.gov)

Also, Roth IRAs have ordering rules that matter if you're pulling money before it's fully "qualified." Publication 590-B explains that distributions are treated as coming first from regular contributions, then conversions/rollovers (FIFO), and only then earnings. (irs.gov)

The point isn't to memorize the ordering rules. The point is to respect that Roth dollars have *options*, and options are valuable in retirement.

7.14 The ``gap years": the most underused opportunity in retirement tax planning

The years after you stop working but before Social Security and RMDs fully kick in can be unusually powerful. Many people see them as a break: "Finally, our income is low."

That's exactly why they're powerful. Low-income years are a chance to decide what you want your future tax picture to look like.

You can use them for:

- Partial Roth conversions (deliberate, not massive).
- Realizing capital gains in years where the tax cost is lower than it would be later.
- Testing your spending plan with real numbers while you still have maximum control over where your income comes from.

The common thread is control. Later, RMDs reduce control. Social Security taxation reduces control. IRMAA cliffs reduce control. The best sequencing happens before control is taken away. (irs.gov)

7.15 Actionable takeaway: build a one-page ``income map" before you take another withdrawal

Set a 45-minute appointment with yourself and create a simple map. Not a spreadsheet masterpiece—a map you can understand when you're tired.

Write down:

- Your expected Social Security start date and annual benefit.
- Your current balances in taxable, tax-deferred, and Roth.

- The year you expect RMDs to begin (based on current law rules) and a rough estimate of how large the tax-deferred account could be by then. (irs.gov)
- Whether your MAGI is likely to flirt with Medicare IRMAA tiers in any year with a conversion, a big sale, or a large withdrawal. (cms.gov)

Then choose one rule you'll follow for the next 12 months:

> *We will withdraw enough from our traditional IRA (or convert to Roth) to reach a planned taxable income target, and we will use taxable or Roth dollars for the rest.*

You're not trying to outsmart the tax code. You're trying to avoid being surprised by it. When your withdrawals are sequenced with intention, the tax torpedo loses its favorite ingredient: a big, unplanned spike.

Chapter 8

The Execution Toolkit

You have the strategy, but a strategy without a system is just a wish. This chapter is about building the engine that drives your financial life so you don't have to push it uphill every day. We're going to move beyond spreadsheets and shoeboxes to set up an automated dashboard that tells you exactly where you stand. We will also lock the doors against the scams targeting your growing nest egg and have a frank conversation about when it's time to stop DIY-ing and hire a pro. This is about removing friction so your plan runs on autopilot.

8.1 Your Digital Financial Dashboard

At 43, Marisol could tell you the balance of her checking account within about $200 at any given moment. She could also tell you, with painful accuracy, that her mortgage payment hit on the first, childcare on the fifteenth, and her credit card autopay "sometime near the end of the month." What she *couldn't* tell you was the number that matters most: whether she was actually getting wealthier.

Her money life had become a scavenger hunt. A retirement account from a previous job. A current 401(k). A savings account she barely touched. Two credit cards. A student loan she was "pretty sure" was on track. Every few months she'd log into everything, squint at a half-dozen dashboards, and walk away with the same conclusion: *"I'm busy, I'm responsible, and I still don't know if this is working."*

That feeling isn't a personal failing. It's an information problem.

A digital financial dashboard is a way to turn your finances from a pile of separate stories into one coherent headline. Not a complicated one. Not a "new hobby." It's more like a speedometer: you don't stare at it all day, but you want it to be accurate when you glance down. The goal is to reduce the number of times you have to ask, "Wait, where did that money go?" and increase the number of times you can say, "I know exactly what's happening, and I'm fine."

The real enemy: cognitive load

People think money problems are math problems. In midlife, they're often attention problems.

When your finances are fragmented, you pay a constant "attention tax." You remember passwords. You chase statements. You mentally reserve money for bills that haven't posted yet. You wonder whether your credit card balance is "real" or just a temporarily ugly number. You keep tiny ledgers in your head: *Okay, if the car insurance hits Friday and the dentist is Monday, then we should be okay as long as…*

That mental spreadsheet is exhausting, and it crowds out better thinking. You can't zoom out to strategy—saving rate, tax choices, investment allocation—because you're stuck zoomed in on whether you accidentally double-paid the water bill.

A dashboard reduces cognitive load by doing one thing relentlessly well: it collects the truth in one place and updates it without requiring your personality to be "meticulous." Midlife is already full. Your system has to work when you're tired.

What your dashboard must do (and what it can ignore)

A useful dashboard has a few non-negotiables:

- *It shows your net worth automatically.* Net worth is simply assets minus debts. It's the scoreboard that tells you whether your daily effort is turning into long-term progress.
- *It captures spending with as little friction as possible.* Not because you need to shame yourself, but because patterns hide in the noise.
- *It separates "cash-flow safety" from "wealth-building."* Checking accounts and credit cards are about *not bouncing*. Retirement accounts and brokerage accounts are about *not running out later*. A dashboard that mixes these together without clarity creates false panic.
- *It creates a small set of repeatable routines.* The point isn't to know everything; it's to know the few things that keep you in control.

A dashboard *does not* need to be perfect. It doesn't need to categorize every transaction flawlessly. It doesn't need to predict the market, optimize every tax angle, or turn you into a monk. It just needs to reduce confusion and make progress visible.

The platform question: app, spreadsheet, or hybrid

There are three common approaches:

1. *Aggregator apps* that link to your financial institutions and pull transactions and balances automatically. (Examples people often use include Monarch, YNAB, and Empower.)
2. *A spreadsheet* you update manually.
3. *A hybrid* where an app does transaction capture and a spreadsheet handles one-off planning (like retirement projections or a debt payoff plan).

The best choice isn't about which tool is "best" in the abstract. It's about which tool you'll actually keep using when your work calendar explodes and your kid gets the flu.

Aggregator apps: the "set it up once" advantage

Aggregators shine at automation. They reduce the friction of gathering data, which is often the biggest barrier to consistency. Most people don't avoid tracking because they hate insight; they avoid tracking because logging in to six places feels like doing taxes as a hobby.

The tradeoff is that apps can be messy at the edges. Accounts occasionally fail to sync. Some transactions get miscategorized. You might need to nudge it back on track now and then. The key is mindset: the app is not a courtroom transcript; it's an instrument panel. You want it directionally accurate and reliably maintained.

Spreadsheets: the "I trust what I touch" advantage

Spreadsheets are flexible and private, and they force you to look at the numbers. For some people, that manual step is the feature, not the bug: it creates a weekly ritual of awareness. They also handle unusual finances well: variable income, side businesses, rental properties, complex reimbursements.

The cost is that spreadsheets demand attention forever. If you fall behind, they turn into an abandoned gym membership: a symbol of good intentions that quietly becomes a guilt object.

Hybrid: automation for the present, spreadsheets for the future

A hybrid approach is often the sweet spot for midlife earners. Let the app handle the daily churn (transactions, balances, categories) and keep a simple spreadsheet for the few planning questions apps don't answer well:

- *How much do I need to save each month to hit my retirement target?*
- *If I pay off this debt first, what happens to cash flow?*
- *What does "good" look like for my emergency fund?*

A planning spreadsheet can be updated monthly or quarterly instead of daily, which keeps it from becoming a second job.

A practical decision table (so you don't overthink it)

If you are...	*Your best starting point*	*Why it works*
Busy, allergic to admin, likely to quit if it's tedious	Aggregator app	Automation keeps the system alive when motivation dips.
Detail-oriented, privacy-focused, or dealing with unusual income/expenses	Spreadsheet (or hybrid)	Manual control builds trust and handles edge cases cleanly.
Motivated but inconsistent; want visibility without obsession	Hybrid	The app captures reality; the spreadsheet answers big questions on a schedule.

Notice what's missing: the hunt for the perfect tool. The "best" system is the one you'll still be using a year from now.

How to build your dashboard in one weekend (without turning it into a personality)

You're building something closer to a kitchen than a museum. Functional beats fancy.

Step 1: Choose your "north star" metrics

Pick a small set of numbers you want to see without effort. A good default set:

- *Net worth* (assets minus debts), updated automatically if possible.
- *Cash runway* (how many months you can cover essential expenses from liquid cash).
- *Savings rate* (the percentage of your income you're keeping, not spending).

That may look like "three things," but each one answers a different kind of fear:

- Net worth answers: *Am I building wealth?*
- Cash runway answers: *Am I safe from a short-term shock?*
- Savings rate answers: *Is my current life compatible with my future?*

If you only track one, track net worth. It's hard to fake. It's also emotionally clarifying: you can have a high income and a stagnant net worth, or a modest income and a net worth that steadily rises. The difference is the gap your dashboard makes visible.

Step 2: Link accounts (or list them) in a way that matches your attention span

If you use an app, link every meaningful account:

- Checking and savings
- Credit cards
- Mortgage and other loans
- Retirement accounts (401(k), IRA, etc.)
- Brokerage accounts

If you use a spreadsheet, create a single tab called `Accounts` with four columns: `Account`, `Type`, `Balance`, `Notes`. Update balances monthly. That's enough to calculate net worth and spot drift.

The trick is not to stop because one account is annoying to connect or track. Get 90% of the picture first. Perfection is a great way to stay stuck.

Step 3: Make categories boring on purpose

Most tracking systems collapse because the categories are too imaginative. People build a taxonomy worthy of a museum archive, then can't decide whether a $37 Target run was "Housing," "Kids," "Groceries," or "Regret."

Keep categories broad:

- Housing
- Food
- Transportation
- Health
- Kids
- Debt payments

- Savings/Investing
- Everything else (yes, really)

You can always refine later, but you'll rarely need to. The purpose of categories is to reveal big patterns, not to win a labeling contest.

Step 4: Turn bills into a predictable rhythm

A dashboard is only calming if your cash flow is calm. If your bills are scattered, your checking account will look like a heart monitor.

Two practical moves help:

- *Put recurring bills on autopay where it's safe to do so.* The goal is fewer "surprise" withdrawals.
- *Align due dates with paydays if possible.* Many companies will change your billing due date if you ask. Even shifting a couple of major bills can smooth your month dramatically.

This isn't about optimization. It's about sleep.

A case study: the dashboard that stopped the ``phantom broke" feeling

Marisol chose an aggregator app because her biggest issue wasn't discipline; it was fragmentation. She linked her accounts and watched the first month of transactions pour in like a hose turned on.

The first week was uncomfortable. Her categories were wrong. The app thought her daycare payments were "Shopping." Her retirement account didn't sync at first. She considered quitting.

Instead, she gave herself one rule: *ten minutes a day for ten days.* Not to perfect it. Just to keep it moving.

By day four she saw something she hadn't expected: her spending wasn't out of control in the dramatic way she feared. It was out of control in a quieter way. Subscriptions, convenience spending, and "small" purchases added up to a number large enough to fund her Roth IRA for the year.

But the bigger surprise was psychological. Once net worth was visible, her money story changed. A big credit card balance used to feel like a verdict: *We're failing.* Now it sat in the same frame as her retirement contributions and mortgage principal. She could see that some months she was still moving forward even when the checking account felt tight.

Then the dashboard exposed the real leak: timing. She was paying the credit card early, then floating essentials on the same card again before payday. Her spending wasn't the problem; her cash-flow sequence was. She shifted due dates where she could, adjusted autopay to a few days after payday, and stopped playing monthly whack-a-mole.

Three months later, she wasn't magically rich. She was *steady.* The phantom-broke feeling faded because she no longer had to guess. She could glance, know, and move on with her life.

That's what a dashboard is supposed to do: not make you money-obsessed, but make you money-*unconfused*.

The weekly check-in that doesn't ruin your weekend

A dashboard only helps if you look at it, but "look at it" doesn't mean "enter a spiral."

A good weekly check-in takes 10–15 minutes and answers four questions:

- *Did anything weird happen?* (a duplicate charge, a forgotten subscription, a medical bill)
- *Do I have enough cash for the next seven days of essentials?*
- *Are my credit cards doing what I think they're doing?* (balances trending down if you're paying them off)
- *Is there one decision to make this week?* (move $200 to savings, pause a subscription, schedule a call to fix a billing error)

If you're partnered, this can be a short conversation, not a summit meeting. The tone matters more than the precision. Treat it like checking the weather before a trip: you're not trying to control the sky, you're trying to dress appropriately.

Net worth: the number that makes grown-up tradeoffs easier

Net worth is blunt, and that's why it's useful.

Income can flatter you. A high salary can mask fragile finances. A low salary can hide impressive discipline. Net worth doesn't care about the story you tell yourself; it reflects the story your money is actually writing.

Tracking net worth monthly also changes how you react to normal life:

- A big home repair becomes a planned draw from cash, not an identity crisis.
- Market volatility becomes a line on a chart, not a personal insult.
- Debt payoff becomes visible progress, not just "still paying."

The trick is to interpret net worth like an adult interprets a scale: one weigh-in is noise, a trendline is information.

Common failure points (and how to avoid them)

Failure point: you quit because the first month looks ugly.

The first month is always ugly. It's not you; it's the data arriving without context. Give it a full cycle of income and bills before you draw conclusions.

Failure point: the system becomes a moral referendum.

A dashboard is not there to judge you. If you feel shame every time you open it, you'll avoid it. Keep categories broad, focus on trends, and treat surprises as clues.

Failure point: you obsess over small expenses and ignore big levers.

A dashboard should *highlight* big levers: savings rate, housing cost, debt interest, insurance premiums, retirement contributions. If you find yourself in a daily argument with your coffee budget while your car loan interest rate quietly burns money, the dashboard has become entertainment instead of tool.

Failure point: you never reconcile anything.

Even automated dashboards need occasional maintenance. A weekly glance catches errors while they're still small. The goal is not perfection; it's preventing drift into nonsense.

A simple build plan you can start tonight

Here's the fastest path to a working dashboard without getting swallowed by it:

1. Pick a platform: aggregator app, spreadsheet, or hybrid.
2. Define your three metrics: net worth, cash runway, savings rate.
3. Gather your account list (everywhere money lives or is owed) and connect or enter them.
4. Create boring categories and promise yourself you won't redesign them for 30 days.
5. Schedule a weekly 15-minute check-in on your calendar for the same day and time.

If you do nothing else, do the calendar step. Most money systems fail because they rely on memory, and memory is busy raising kids, doing work, and trying to stay healthy.

Actionable takeaway: your first dashboard snapshot

Before you close this book tonight, take 12 minutes and create a "snapshot" note in your phone or a document called `Dashboard Baseline`. Write down:

- Your checking balance (right now)
- Your total credit card balance (right now)

- Your retirement account total (even a rough number)
- Your mortgage balance (or rent, if you rent)
- One sentence: *"If I checked these numbers weekly, I would feel..."*

That snapshot is the seed of a dashboard. It turns vague anxiety into a defined starting line, and it proves something important: clarity doesn't require a perfect system. It requires a system you'll actually return to.

8.2 Protecting Your Wealth from Fraud

Fraud is the only "expense" that can hit your life like a meteor. It doesn't arrive politely at the end of the month. It doesn't ask if your cash flow is ready. It just lands: a drained checking account, a new credit card you didn't open, a wire transfer you didn't authorize, a voicemail that sounds like your boss, a text that looks exactly like your bank.

And if you're in your 40s or 50s, you're in a particularly tempting slice of the target market. You've likely accumulated some assets. You may have a stable income. You probably have more accounts than you did at 28 (old 401(k)s, a couple of cards, a mortgage, maybe a brokerage account). You might be managing not only your own money but also a parent's logistics or a teenager's first bank account. Complexity is oxygen for fraud.

The goal here isn't to become paranoid. It's to become *boring to attack*. Professional scammers look for speed, confusion, and high trust. When your defenses are layered and your accounts are harder to hijack, they tend to move on.

A simple truth: awareness is not a security system

Most people's anti-fraud plan is basically a vibe: "I'm pretty careful." They don't click obvious spam. They shred some mail. They assume that if something bad happens, their bank will fix it.

That "careful" feeling is comforting, but it's not a control. Controls are settings you turn on and habits you install so you don't have to rely on being sharp every single day. Midlife is busy. Your defenses should keep working when you're tired, distracted, or dealing with real life.

The difference matters because modern fraud often doesn't look like a cartoonish Nigerian prince. It looks like a calm, plausible request delivered at the exact moment you don't have time to think.

The midlife scam menu (and why it works)

Scammers don't need to fool everyone. They just need to fool someone who is rushing.

Imposter fraud is the workhorse: someone claims to be your bank, the IRS, Social Security, a delivery company, your employer, or a tech support agent. The message is designed to create urgency: suspicious activity, frozen account, missed payment, legal threat, "final notice."

Account takeover is quieter: a stolen password from an old data breach gets tried across financial sites. If you reuse passwords (many people do), a scammer may not need to persuade you of anything at all.

Investment and "recovery" scams are the cruel ones. The pitch is either easy money (crypto, real estate flips, private deals that are "not public yet") or getting your money back after a loss. If you've already been scammed once, a "recovery agent" may contact you and scam you again.

Family-angle scams spike with emotion: "grandparent" scams, hacked social media accounts, and AI-generated voice messages that sound like someone you love. The request is simple: send money now, don't tell anyone, I'm embarrassed.

All of these rely on the same mechanics: get you to act before you verify. The best defense is a system that makes "verify" automatic.

Layered security: build a moat, then a gate, then a wall

A useful way to think about protection is layered defense. If one layer fails, the next one is still standing. You don't need twenty fancy tools; you need a few high-impact layers that cover the common paths criminals use.

The three layers that do the most work for the least effort are:

- a password manager (so your passwords can be strong and unique without living in your head),
- multi-factor authentication (so a stolen password isn't enough),
- a credit freeze (so identity thieves can't easily open new credit in your name).

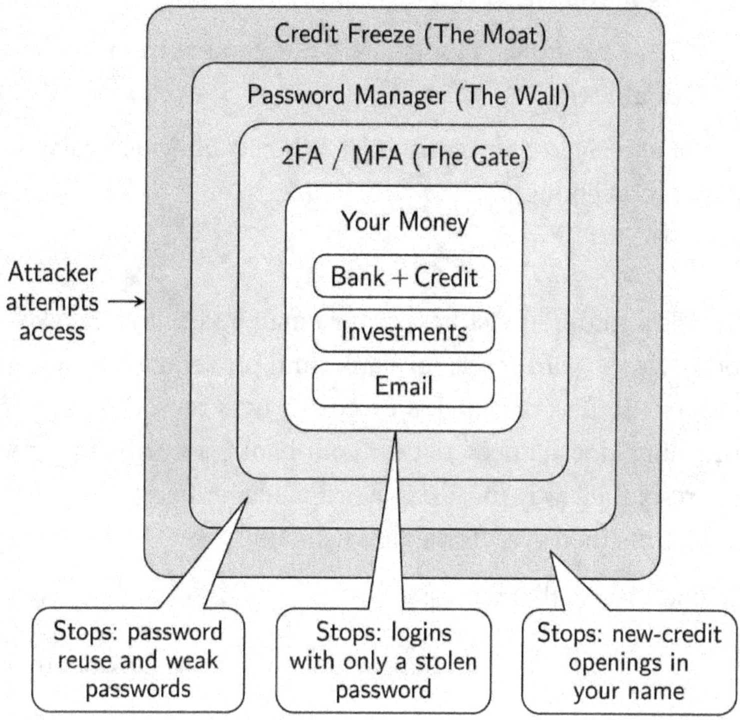

Layer 1: Password manager (the wall)

A password manager is an app that stores your passwords in an encrypted vault so you don't have to remember them. You remember one strong master password (and ideally protect it with multi-factor authentication), and the manager generates unique, long passwords for everything else.

This single change quietly solves three problems that fuel most account takeovers:

- *Password reuse.* If one site leaks your password, criminals try it everywhere.
- *Weak passwords.* Humans are not good at inventing randomness.
- *Convenience overrides caution.* When you're busy, you pick "good enough."

There's a midlife twist here: your email account is the skeleton key. Password resets for banks and brokerages often start with email. If your email is protected by a reused password, a criminal doesn't need to beat your bank's security; they can just reset their way in. A password manager helps you fortify email first, then everything downstream.

Practical setup that works:

- Put your password manager behind multi-factor authentication.
- Change your email password to a long, unique one generated by the manager.
- Update your primary financial logins next: bank, credit cards, brokerage, retirement plan, and any account that can move money.

The best day to do this was years ago. The second-best day is a Tuesday night when you're tired of carrying your financial life in your head.

Layer 2: Multi-factor authentication (the gate)

Multi-factor authentication (MFA, sometimes called 2FA) means logging in requires two things, not one. The classic example is a password (something you know) plus a one-time code (something you have). If a scammer steals your password, they still can't get in without the second factor.

Not all second factors are equal.

Text-message codes (*SMS*) are common and better than nothing, but they're not the strongest choice. Criminals can sometimes pull off "SIM swap" attacks (convincing a phone carrier to move your number to a new SIM), intercepting those codes.

Authenticator apps (time-based codes) are generally stronger than SMS. They still require careful backup planning so you don't lock yourself out if your phone is lost.

Hardware security keys (like a YubiKey) are among the strongest options for critical accounts that support them, because the "key" is a physical device you tap or insert. They also reduce the risk of being tricked by a fake login page, because the key won't authenticate to the wrong site.

You don't need hardware keys for your grocery store rewards account. You want them—or at least app-based MFA—for the accounts that can cause catastrophic damage:

- email accounts,
- bank and credit card accounts,
- brokerage and retirement accounts,
- phone carrier account (yes, really),

- password manager itself.

A small but powerful habit: when you get an MFA prompt you didn't initiate, treat it like a smoke alarm. Don't "approve to make it stop." Deny it, change your password, and investigate.

Layer 3: Credit freeze (the moat)

A credit freeze is one of the highest-leverage identity theft defenses available because it blocks a common move: opening new credit in your name. When your credit files are frozen, most lenders can't pull your credit report, and without that, many won't open a new account.

It's also free to place and free to lift, and it doesn't affect your credit score.

There's a practical catch: you have to freeze with each major credit bureau (Equifax, Experian, and TransUnion) individually. And later, when you legitimately need new credit (a mortgage refinance, a new credit card, sometimes a utility account), you temporarily lift the freeze with the specific bureau the lender uses. That adds a small step, but it also adds a powerful "speed bump" for criminals.

If you want a system, not a project:

- Freeze all three bureaus.
- Store your freeze PINs or credentials in your password manager.
- When you need credit, lift the freeze for a short window (for example, a few days), then refreeze.

The peace of mind is disproportionate to the effort. It's one of the few security moves that meaningfully reduces risk without requiring you to be vigilant every day.

Case study: the ``bank text" that almost worked

Darren, 49, got a text while in a meeting: "Bank Alert: Suspicious charge detected. Reply YES to confirm or NO to deny." It looked right. Same tone, same short links his bank had used before. His heart rate did that instant thing it does when money feels threatened.

He replied `NO`. Immediately another message arrived: "Your account has been locked. To restore access, verify your identity:" followed by a link.

He clicked.

The page asked him to log in. He did. Then it asked for the one-time code texted to his phone. He entered that too, because the whole point was to "restore access."

Within minutes, the scammer used Darren's credentials and the fresh MFA code to log in for real and add a new transfer recipient. The scammer tried to move money out.

Two things saved him. First, his bank flagged the transfer as unusual and delayed it. Second, Darren's weekly dashboard habit meant he noticed quickly. He called the bank using the number on the back of his card (not the one in the text) and stopped the transfer.

When Darren cleaned up afterward, the lesson wasn't "I'm an idiot." The lesson was that the scam was designed to hijack the *exact* safety behavior conscientious people perform: responding quickly to an alert.

He made three changes that week:

- He turned on app-based MFA wherever possible and tightened his email security.
- He moved his passwords into a manager and made them unique.
- He created a personal rule: never click a link in a financial text; open the bank app directly or type the website manually.

The goal isn't to never be fooled by a clever lie. The goal is to make "being fooled" less expensive.

The habits that keep your layers from cracking

Security is part settings, part rhythm. A few habits are worth installing:

Use a "known-good" contact method. If a message claims to be your bank or broker, don't continue the conversation inside the message. Use the app, or call the number on your card, or pull the number from an old statement. This one habit kills an enormous percentage of imposter scams.

Treat urgency as a red flag, not a cue. Real institutions can be urgent; scammers weaponize urgency. When you feel pressured, slow down. That feeling is the hook.

Keep your phone carrier account locked down. If your phone number gets hijacked, a scammer can intercept codes and password resets. Use a strong password and any extra account protections your carrier offers.

Turn on transaction alerts for key accounts. A small notification for large transfers, new payees, or card-not-present charges can turn "months later" into "caught it today."

Check credit reports on a schedule. You can access legally available free credit reports through the official channel at AnnualCreditReport.com. The point isn't to stare at them weekly; it's to make sure no new accounts appear without your consent.

If something goes wrong: a calm, fast response plan

Fraud feels personal, but the response should be procedural.

- *Stop the bleeding:* contact the financial institution using a known-good number, freeze cards, change passwords, remove unknown payees, halt transfers.
- *Secure the keys:* lock down email first, then your password manager, then financial accounts.
- *Document:* dates, amounts, screenshots, who you spoke with.
- *Use the official identity theft recovery pathway:* IdentityTheft.gov can generate a tailored recovery plan and help you track steps.

A final, important note: scammers sometimes impersonate the very agencies meant to help (including "FTC" impersonation). That's another reason to use known-good websites and numbers rather than trusting inbound calls or texts.

Actionable takeaway: a 30-minute ``castle build" sprint

Put a timer on for 30 minutes and do the highest-impact sequence:

1. Enable MFA on your primary email account (prefer an authenticator app or hardware key if available).
2. Change your email password to a long, unique one.
3. Turn on MFA for your bank and your main credit card login.
4. Place a credit freeze with Equifax, Experian, and TransUnion, and store the login details/PINs in your password manager (or a secure place you can actually find later).

When that's done, you've made yourself dramatically harder to rob. Not invincible. Just inconvenient—and inconvenience is one of the most underrated forms of wealth protection.

8.3 Building Your Financial Team

Hiring financial help in midlife can feel like walking into a car dealership when you don't know the sticker price. Everyone is friendly. Everyone is confident. Everyone has a pen. And somehow the conversation slides from *advice* to *products* before you've even said what you want.

The uncomfortable truth is that "financial advisor" is a job title the way "health expert" is a title. It might mean a skilled professional who can save you tens of thousands of dollars in mistakes. It might mean a salesperson with a quota and a glossy brochure. The difference isn't vibe. It's incentives, legal duties, and whether their recommendations are built for *your* life or for their commission statement.

A strong dashboard gives you visibility. Strong security keeps your money from disappearing. A strong team keeps you from making expensive decisions in the dark—or paying for help that quietly makes you poorer.

8.4 The first rule: stop shopping for a person, start shopping for a job

Most people begin with, "Do I need an advisor?" That's like asking, "Do I need a contractor?" Maybe. For what?

Financial help is not one job. It's a set of jobs that sometimes overlap:

- *Tax planning and tax filing:* a CPA or enrolled agent can help you file correctly, find deductions and credits you're entitled to, and plan for big moves (like stock compensation, a business sale, or a rental property).
- *Retirement planning and investment advice:* a financial planner can help you set goals, choose an investment strategy, and coordinate accounts, insurance, and taxes.
- *Estate planning:* an estate attorney can draft documents (wills, trusts, powers of attorney) that keep your family from a legal mess and reduce the chance of conflict or exploitation.

- *Insurance analysis:* sometimes done by a planner, sometimes by an agent (with sales incentives), often best approached with extra skepticism.

You don't hire "a finance person" the way you hire a babysitter. You hire expertise to solve a specific problem with money stakes attached.

A surprisingly effective starting question is: *What decision am I avoiding because it feels too consequential?* That decision is your job posting.

8.5 When DIY is smart (and when it becomes expensive pride)

There's a certain badge of honor in doing it all yourself. It's also, occasionally, a trap.

DIY works well when your financial life is simple and stable:

- one or two incomes,
- a straightforward retirement plan at work,
- no business ownership,
- no stock compensation,
- no complex tax situations,
- and a willingness to learn.

DIY becomes risky when complexity creeps in, especially when the cost of a mistake is large and irreversible. Midlife is exactly when those situations show up: career changes, inheritances, divorce, blended families, caring for parents, kids approaching college, a home purchase or sale, a business taking off, or a sudden jump in income with unfamiliar tax consequences.

The painful part is that complexity doesn't announce itself. It whispers. You realize you needed help *after* the mistake, the way you realize you needed a lawyer after you've already signed.

So the question isn't "Am I smart enough?" It's "Is the downside of being wrong big enough to justify paying for expertise?"

8.6 The core distinction: fiduciary vs. salesperson

A fiduciary is someone who is legally obligated to put your interests first. That sounds like the baseline, but it isn't universal in the financial world.

Many people who call themselves "advisors" are compensated by selling products: mutual funds with sales loads, annuities, life insurance policies, or other investments that pay commissions. They may also charge fees layered on top. Some are excellent and ethical. Some are not. The problem is that you can't reliably tell which is which from charisma.

A fiduciary's edge isn't that they're nicer. It's that their legal duty and compensation structure can be aligned with advice—not sales.

Two phrases worth learning:

- *Fee-only:* the advisor is paid by you (hourly, flat fee, or a percentage of assets they manage) and does not receive commissions for selling products.
- *Fee-based:* sounds similar, but often means they charge fees *and* can earn commissions. That "and" is where conflicts live.

There are good advisors in both camps, but "fee-only fiduciary" is a useful filter if your goal is to reduce hidden incentives. When you're paying for advice, you want your money buying thinking, not inventory.

8.7 How to spot incentives in the wild (without becoming cynical)

Incentives aren't always evil. They're just physics. People move toward what pays them.

If someone is compensated by selling life insurance, they will see life insurance solutions everywhere—in the same way a roofer will notice roofs. That doesn't mean roofs aren't important. It means you must understand what the person across the table gets paid to notice.

Here's the simplest way to bring incentives into the open: ask, early and calmly,

> "How do you get paid, in plain language? If I follow your recommendation, who else pays you?"

A professional who is used to working transparently will answer directly. Someone who deflects, minimizes, or treats the question as rude is giving you a preview of how future conflicts will go.

Another useful tell: listen for the moment the conversation shifts from your goals to a specific product. Advice should feel like a map. Sales feels like a shelf.

8.8 The team you actually need (and the moments you need them)

In your 40s and 50s, the most valuable financial help often looks less like stock picking and more like *coordination*. The work is connecting decisions that interact:

- Retirement contributions affect taxes.
- Taxes affect cash flow.
- Cash flow affects debt payoff.
- Debt payoff affects how much risk you can take.
- Insurance and estate documents protect the whole structure.

A good team doesn't just answer questions; it prevents you from asking the wrong questions.

Here are common "hire a pro" triggers:

You have stock compensation (RSUs, options) and your income is lumpy. This is a classic place where tax mistakes are easy and expensive.

You're self-employed or own part of a business. Retirement plan choices alone can change your long-term outcome dramatically, and taxes get more nuanced.

You're entering or exiting a marriage, or managing a blended family. This isn't just emotional complexity. It's beneficiary designations, titling, legal documents, and the difference between "what we intended" and "what happens."

You're caring for a parent or receiving an inheritance. Decisions come with family dynamics, legal timelines, and the risk of exploitation.

You've accumulated multiple retirement accounts and you're not sure they work together. Consolidation, investment allocation, and beneficiary setup can be simplified with competent guidance.

You're close enough to retirement that mistakes don't have time to heal. Time is the great eraser when you're 25. At 55, less so.

8.9 A case study: the advisor who "found" $90,000 (without higher returns)

Sam and Priya were 47 and 45, with two kids, good incomes, and that familiar midlife feeling: *We're doing okay, but are we doing enough?*

They met with an advisor through a referral. The first meeting felt great. The advisor was confident and warm. He pointed at their accounts and said something like, "You're leaving money on the table." Then he recommended moving their retirement accounts into products he managed, plus a cash-value life insurance policy as a "tax-free retirement vehicle."

Sam almost signed. The pitch was persuasive: protect the family, build wealth, simplify everything.

Priya hesitated and asked two questions:

- "Are you a fiduciary at all times?"
- "How do you get paid?"

The answer was slippery. He said he operated under "best interest standards" and that his compensation "varied." He didn't say "commission" unless prompted.

They paused and did something most people don't: they verified. They used public lookup tools to review licensing and disclosures and realized the advisor's primary business was selling insurance products.

Instead of giving up, they hired a fee-only planner for an hourly engagement—a clear scope: review retirement contributions, taxes, and insurance needs.

The planner didn't promise higher returns. She didn't talk about beating the market. She did something more valuable: she cleaned up the plumbing.

- She noticed Sam was contributing to the 401(k) but missing the full employer match because contributions weren't timed correctly across the year.
- She helped them adjust withholding so they weren't unintentionally giving the IRS an interest-free loan, freeing up cash flow.

- She pointed out that their term life coverage was too low for their actual obligations and recommended straightforward term insurance rather than a complex policy.
- She reviewed their old retirement accounts and suggested consolidating where it made sense to reduce fees and simplify investment allocation.

Between capturing the full match, reducing avoidable fees, and improving cash flow management, they estimated the difference over the next decade could exceed $90,000—not because she picked magical investments, but because she removed friction and leakage. It felt almost unfair: the "sexy" advice was the one that paid someone else. The "boring" advice was the one that paid them.

8.10 How to verify a professional (without trusting Google reviews)

Midlife earners are often targeted precisely because they're competent and busy. They'll do due diligence on a contractor, but they'll trust a financial professional because the suit looks expensive.

Verification is a skill. It's also respectful: good professionals expect it.

Here are the checks that matter:

- *Investment Adviser Public Disclosure* (*IAPD*) / *SEC:* for registered investment advisers (RIAs). You can see registration status and disclosures.

- *FINRA BrokerCheck:* for brokers and brokerage firms. It shows licensing and disclosures drawn from the industry registration database.
- *CFP Board verification:* if someone claims to be a CFP® professional, you can confirm certification and view certain public disciplinary information.
- *IRS preparer resources:* if you're hiring a tax preparer, use IRS guidance and directories for certain credentialed preparers, and be cautious about anyone promising unusually large refunds.

If that sounds like a lot, remember the point: you're not trying to become a regulator. You're trying to confirm that the person is who they say they are, with the credentials they claim, and without a trail of serious disciplinary events you'd want to know about.

A quiet benefit: a professional who knows you'll verify tends to stay on their best behavior.

8.11 The compensation menu (and what you're really paying for)

Understanding compensation keeps you from comparing apples to smoke.

Common models:

Hourly or project-based planning. You pay for a defined scope: review retirement plan, build a cash-flow plan, analyze insurance, create a college strategy, or run a retirement projection.

This can be ideal when you want a second brain but not ongoing management.

Retainer. You pay a monthly or annual fee for ongoing access and periodic planning updates. This can make sense if your life is complex and changes frequently.

Assets under management (*AUM*). The advisor charges a percentage of the money they manage (often around 1% as a rough reference point, sometimes lower at higher asset levels). This model can align incentives with growth, but it can also be expensive and may create incentives to keep assets under management rather than paying down debt or buying a home.

Commission-based. The advisor is paid when you buy a product. Sometimes this is appropriate for certain insurance needs; it can also create pressure toward complexity and higher-cost products.

None of these are automatically "good" or "bad." The key is fit:

- If you want a plan and you'll execute it yourself, hourly/project is often the cleanest.
- If you want ongoing coordination and behavior support, retainer may be worth it.
- If you want hands-off investment management and value delegation, AUM can work—but you should understand the dollar cost and what you're getting beyond portfolio rebalancing.

Ask for the fee schedule in writing. Then do the math in dollars, not percentages. One percent sounds tiny until it's $6,000 a year, every year, compounding in the wrong direction.

8.12 The interview questions that reveal everything

A good interview feels less like being sold to and more like being diagnosed.

Questions that tend to separate professionals from performers:

- "Are you a fiduciary at all times? If not, when are you not?"
- "How do you get paid? Please include commissions, referral fees, and revenue sharing."
- "What is your typical client situation, and what problems do you solve most often?"
- "Do you provide a written scope of work and a written summary of recommendations?"
- "What do you *not* do?" (Specialists have edges; generalists have limits.)
- "If I hire you, what decisions will I be able to make more confidently within 90 days?"

Listen not just to the answers, but to the *texture*. Clear, direct answers suggest a professional who works in daylight. Defensive answers suggest someone who benefits from fog.

8.13 Red flags that deserve an immediate exit

Some warning signs are subtle. Others are a fire alarm.

Walk away if you encounter:

- guaranteed returns or "can't lose" language,
- pressure to act quickly or sign immediately,
- reluctance to provide fees in writing,
- refusal to explain a product in plain language,
- discouraging you from involving your spouse/partner or another trusted person,
- recommending moving everything before understanding your full picture.

Also be cautious with any strategy that requires you to not quite understand it. If the plan depends on your confusion, it's not your plan.

8.14 How to use a professional without outsourcing your life

The best relationships are collaborative. You don't want to hand over your brain; you want to rent expertise.

A simple way to stay in the driver's seat:

- Keep your dashboard active so you always know the basics.
- Ask for a one-page summary after meetings: what changes, why, and what to watch.

- Require plain-language explanations for any product or account move.
- Treat recommendations as hypotheses that must be justified, not commandments.

This approach also protects you from a common midlife failure mode: paying for advice you never implement. Execution is where plans go to die. Your job is to make sure the work turns into actions you can actually sustain.

8.15 Actionable takeaway: write a one-paragraph ``job posting'' for your money

Open a blank document and write a short paragraph titled `Financial Help Job Posting`. Include:

- your top two money stressors (specific, not philosophical),
- the next big decision you need to make (within 6–12 months),
- the type of help you want (hourly plan, tax planning, estate documents, ongoing management),
- and your non-negotiables ("fiduciary at all times," "fee-only," "written scope," etc.).

Then send that paragraph to two or three candidates and see who responds like a professional: clarifying questions, a clear scope, transparent fees, and a calm explanation of what they can do for you. The right team doesn't make you feel dazzled. It makes you feel steadily, quietly more in control.

8.16 The Road Ahead

There's a particular kind of panic that shows up around 40, 45, 52. It's not the dramatic kind where you run outside and wave your arms. It's quieter: you open a retirement calculator, type in a number you *think* is reasonable, and the result stares back like a verdict. The instinctive response is to sprint. To fix everything this year. To become a new person by Monday.

That urge is understandable. It's also the most common way people burn out.

The most valuable shift in midlife isn't learning a new investing trick. It's swapping the mental model from *catching up* to *staying consistent*. Catching up is a mood. Consistency is a system. Catching up is adrenaline and late-night spreadsheets. Consistency is a boring Tuesday where you do one small thing that keeps your future from slipping out of your hands.

And the strange part? The boring Tuesday is what makes you rich.

The decade that decides things (and why intensity lies)

In your 20s and 30s, money mistakes feel recoverable because time is still doing most of the work. In your 40s and 50s, time becomes more like a teammate you need to coordinate with. You can still win. You just can't rely on chaos.

Intensity is seductive because it looks like progress. You cancel every subscription, sell half your closet, swear off restaurants, move money around three times in a week, and tell yourself you're finally serious. For a month, the graphs look great. Then life happens: a child needs braces, a parent's health changes, your company reorganizes, your energy dips, and your intense plan turns into a plan you avoid opening.

Consistency doesn't feel like progress in the moment because it's deliberately unheroic. It chooses small actions that you can do when you're tired, annoyed, or busy. It's what you can maintain when you're not inspired. That's the whole point.

Consistency also scales. When your income rises, or your mortgage drops off, or your kids leave the payroll, a consistent system automatically captures the improvement instead of letting it evaporate into "we deserve it" spending. And when income falls, consistency creates early warnings before the problem becomes a crisis.

A good financial life after 40 has the same vibe as a well-run kitchen: a few sharp knives, everything in reach, and a routine that makes dinner possible even on a long day. You don't need a new personality. You need a counter you can keep clear.

A real-world case: the couple who stopped trying to be perfect

Consider Marcus and Elena, both 47, two kids, suburban life, decent jobs, and the classic feeling that everyone else got the memo earlier. Marcus had a 401(k) he contributed to sporadically. Elena had an IRA she meant to max out every year but usually didn't. They weren't reckless; they were *inconsistent*. Every financial decision felt like a tiny argument with the calendar: "We'll do it after this work project," "after soccer season," "after the holidays," "after the bonus."

They tried the classic sprint. In January, they built a strict budget and swore they'd follow it exactly. February went okay. March had a car repair and a school trip. The budget blew up. By April, they stopped tracking entirely, because tracking felt like admitting defeat.

The change wasn't a new app, a new portfolio, or a new guru. The change was agreeing on one sentence: *We will be the kind of people who review money on purpose, even when it's messy.*

They picked a cadence that didn't require motivational fireworks:

- A 20-minute money check-in on the first Sunday of each month.
- A quarterly "goal check" to decide whether their current savings rate still matched their real life.
- An annual "defense day" for insurance, beneficiaries, and fraud prevention.

Then they made the rules kind, not cruel. Their monthly check-in had exactly three questions written on an index card:

1. What changed since last month?
2. What do we need to do once?
3. What do we need to automate or simplify?

They weren't hunting for perfect categories. They were hunting for leaks, drift, and forgotten tasks. Within six months, their "messy but regular" routine outperformed their "perfect but abandoned" plan. Not because they became disciplined superheroes, but because they stopped relying on willpower as their primary financial strategy.

The lesson is uncomfortable and freeing: you don't rise to the level of your intentions. You fall to the level of your default settings.

The boring middle: where plans go to die (or quietly work)

Most financial advice is built for two moments: the wake-up call and the victory lap. The wake-up call is when you feel behind and want a reset. The victory lap is when you're close enough to retirement that every contribution feels dramatic.

The tricky part is the middle years—the stretch where you're doing the right things, but the finish line is still far enough away that it doesn't emotionally reward you. Markets go up and down. Your life gets more complex, not less. The temptation is to keep redesigning the plan so it feels new again.

A system that survives the boring middle has two qualities:

1. It's *repeatable*. You can do it without a surge of energy.
2. It's *forgiving*. When you miss a month, you resume without shame spirals.

That's why a cadence matters more than a perfect forecast. Your future will be shaped less by whether you guessed an exact retirement date and more by whether you keep showing up to your own life with regularity.

Here's a useful reframe: the goal isn't to "be good with money." The goal is to make a handful of financial actions as normal as brushing your teeth.

A cadence that keeps you safe and moving

A sustainable cadence is not a high-performance calendar. It's a checklist you trust. Think in layers: monthly for cash flow, quarterly for direction, annually for protection.

Monthly: cash-flow reality check. Not a forensic audit. A reality check. You're looking for: (1) any spending that jumped unexpectedly, (2) any bills you forgot were annual or semiannual, (3) whether your buffer (cash reserve) is still doing its job. If you find nothing, that's not wasted time. That's proof your system is stable.

Quarterly: goal check and drift correction. This is where you ask whether the story of your money still matches the story of your life. A quarterly check is also a clean moment to do light investing maintenance. For many people, rebalancing (bringing investments back toward a target mix) works best when it's infrequent and rule-based: either time-based (every 6–12 months) or threshold-based (only when an allocation drifts by a set percentage). The core idea is to avoid constant tinkering while still preventing the portfolio from wandering into a risk level you didn't choose.

Annually: defense day. This is the one that feels least urgent right up until the moment it becomes urgent. Insurance, beneficiaries, estate documents, credit freezes, device security, and fraud monitoring are not "set it and forget it." They're maintenance. One day a year keeps your financial life from being undone by a single bad week.

If you've ever said, "I'll handle that later," you already understand why this works. You're not lazy. You're human. A cadence is how you stop "later" from stealing your attention.

The psychology of consistency: reduce decisions, increase defaults

Consistency isn't built by trying harder. It's built by deciding less.

Every repeated decision creates a doorway for fatigue. Should we invest more this month? Should we pay extra on the mortgage? Should we cut spending now or wait? When you have to re-argue your plan with yourself every time, the loudest emotion wins. Sometimes the loudest emotion is fear. Sometimes it's exhaustion. Sometimes it's "we deserve a break."

A well-designed system moves the "should we?" decisions into a scheduled review, and leaves defaults in place the rest of the time. Defaults are powerful because they work on normal days.

Three defaults worth earning:

1. *Automatic savings that rises with income.* If you get a raise, a portion automatically goes to your future self before your lifestyle absorbs it.
2. *A clear spending floor.* You know the handful of categories where you refuse to live miserably (maybe travel, maybe good groceries, maybe a hobby that keeps you sane). This prevents the "all sacrifice" cycle that leads to rebellion spending later.

3. *A rule for windfalls.* Bonus, tax refund, inheritance, stock vest, side gig spike. Decide in advance what percentage goes to goals (debt, investing, sinking funds) versus fun. Windfalls are where consistency often breaks because they feel like exceptions. Build them into the plan.

Defaults are not about being rigid. They're about making the right thing the easy thing.

Progress you can feel: milestones that don't sabotage you

A common problem after 40 is that the only "milestone" people recognize is the big one: "retire." That's like trying to stay fit by only celebrating when you run a marathon. The gap between daily effort and distant reward is too wide; motivation falls into it.

You need smaller markers that are emotionally satisfying and financially intelligent. The trick is to celebrate in ways that don't reverse the win.

Examples that work because they lock in progress:

- Funding a full month of expenses in a cash buffer, then the second month, then the third.
- Paying off a specific debt and redirecting that payment automatically into investing the next month (no "found money" period).
- Hitting a retirement contribution target for the year, then using a *planned* treat that fits the budget.
- Completing "defense day" and knowing your financial life is harder to hack, steal, or derail.

Celebration matters because this is not purely mathematical. It's behavioral. A plan you resent will eventually get abandoned. A plan that occasionally rewards you—on purpose—is more likely to survive the long stretch.

How to handle setbacks without resetting your identity

Setbacks are not evidence that you're bad with money. They're evidence that you're alive.

The danger isn't the setback itself; it's the meaning you assign to it. People make a 1,800$mistakeandrespondwitha$18,000 resignation. "We blew it, so why bother?" That thought is expensive.

A consistent system treats setbacks like weather:

1. *Name it:* what happened, precisely?
2. *Contain it:* what's the smallest action that prevents it from spreading?
3. *Adjust:* what rule or automation would reduce the chance of a repeat?

For Marcus and Elena, the car repair that used to trigger a budget collapse became a routine transfer from a "car fund" they built slowly. Not because they were smarter than before, but because they stopped pretending predictable surprises were rare. A predictable surprise is just a bill you haven't admitted yet.

The real flex after 40 is not never making a mistake. It's recovering quickly, with minimal drama, and without punishing yourself into avoidance.

A consistent investor is often a better investor

At this stage, your investing results will be shaped by a few big behaviors more than by cleverness:

- Staying invested through downturns.
- Keeping costs and taxes reasonable.
- Taking an amount of risk you can actually live with.
- Rebalancing occasionally, not obsessively.

The market rewards patience and punishes panic. That's not motivational poster talk; it's the basic reality that selling in fear locks in losses and often misses the rebound.

If you're prone to anxiety, your "best" portfolio on paper may be the one you can't stick with. A slightly more conservative mix that you hold consistently can outperform an aggressive mix you abandon at the worst moment. The right portfolio is the one that survives your nervous system.

That's also why infrequent, rule-based rebalancing is so effective: it gives you a reason to act that isn't emotion. Either it's time (every 6 or 12 months), or the portfolio drifted beyond a boundary you chose in calm weather. You don't need to outsmart the market. You need to outlast your own worst instincts.

When consistency gets hard: life transitions and "financial gravity"

After 40, life transitions create financial gravity. Money starts moving toward new obligations unless you actively direct it.

Common gravity wells:

- Helping aging parents.
- Higher healthcare costs.
- Kids becoming more expensive right before they become independent.
- Divorce or blending families.
- Career shifts, layoffs, or burnout-driven changes.

None of these are moral failures. They're realities. Consistency doesn't mean pretending life is stable; it means building a plan that expects instability.

Two practical moves help:

1. *Keep your plan modular.* Separate your finances into clear buckets (monthly bills, short-term reserves, long-term investing). When a transition hits, you can adjust one bucket without tearing up the whole system.
2. *Update your "enough" number occasionally.* Not daily, not weekly. Occasionally. Life changes, and pretending your old targets still apply is how people drift into quiet panic.

The goal is not to control everything. The goal is to stay oriented when everything is moving.

A simple promise that keeps the decade on track

If you want one promise that's strong enough to hold you through the next decade, make it this:

Once a month, I will face the numbers with curiosity instead of judgment.

Curiosity keeps you in the game. Judgment makes you flinch. Curiosity asks, "What's actually happening?" Judgment says, "I'm terrible at this," which is the fastest route to avoidance.

Facing the numbers doesn't mean you like them. It means you refuse to be surprised by them. It means you let reality be the starting point, not the enemy.

A decade from now, the biggest difference won't be whether you found the perfect tool or the perfect rule. It will be whether you built a rhythm you could live with—a rhythm that quietly collected wins, caught problems early, and made the future feel less like a threat and more like a place you're deliberately funding.

And when motivation fades, as it always does, you'll still have the most powerful financial advantage available to an ordinary person with a busy life: a plan that runs even when you don't feel like it.

www.ingramcontent.com/pod-product-compliance
Lightning Source LLC
LaVergne TN
LVHW052021200926
848481LV00005B/1550